PERSPECTIVES ON MATHEMATICAL PRACTICES

LOGIC, EPISTEMOLOGY, AND THE UNITY OF SCIENCE

VOLUME 5

Logic, Epistemology, and the Unity of Science aims to reconsider the question of the unity of science in light of recent developments in logic. At present, no single logical, semantical or methodological framework dominates the philosophy of science. However, the editors of this series believe that formal techniques like, for example, independence friendly logic, dialogical logics, multimodal logics, game theoretic semantics and linear logics, have the potential to cast new light no basic issues in the discussion of the unity of science.

This series provides a venue where philosophers and logicians can apply specific technical insights to fundamental philosophical problems. While the series is open to a wide variety of perspectives, including the study and analysis of argumentation and the critical discussion of the relationship between logic and the philosophy of science, the aim is to provide an integrated picture of the scientific enterprise in all its diversity.

Perspectives on Mathematical Practices

Bringing Together Philosophy of Mathematics, Sociology of Mathematics, and Mathematics Education

Edited by

Bart Van Kerkhove and Jean Paul van Bendegem
Vrije Universiteit Brussel, Belgium

Springer

A C.I.P. Catalogue record for this book is available from the Library of Congress.

ISBN-10 1-4020-5033-X (HB)
ISBN-13 978-1-4020-5033-6 (HB)
ISBN-10 1-4020-5034-8 (e-book)
ISBN-13 978-1-4020-5033-6 (e-book)

Published by Springer,
P.O. Box 17, 3300 AA Dordrecht, The Netherlands.

www.springer.com

Printed on acid-free paper

Contents

Introduction

Is mathematics finally going through the Kuhnian revolution that the sciences or, more precisely, the philosophers, historians, sociologists, economists, psychologists of science, … have been able to deal with ever since the magical year of 1962? Apart from the fact that one cannot easily identify a book that has played the part that *The Structure* has played – of course, Lakatos' *Proofs and Refutations* comes pretty close, but it does not possess the generality of Kuhn's work – there seems to be plenty of reasons why mathematicians and philosophers of mathematics are reluctant to cheer the coming of a Kuhnian revolution in their favourite domain. Instead of a full-fledged historical-philosophical analysis (actually, many papers in this volume do precisely that, so it is quite unnecessary to duplicate their efforts in this introduction), let us just repeat once more the overused quote: "Mathematics is a free creation of the human spirit". In a nutshell it expresses the cherished beliefs that many share: mathematics stands on its own, free from any societal influence, individualist and immaterial, beyond space and time, in short, it occupies a universe of its own. This view usually, though not necessarily, goes together with a belief, if not a conviction, that mathematical capacities are innate, i.e., one is born a mathematician and a mathematical training merely serves to refine the powers already present. One just needs to remind oneself of the well-known story told by G. Hardy about his reluctance to familiarise the celebrated Indian mathematician Ramanujan with the notion of a proof in mathematics for fear of ruining his innate capabilities. Add to this that to a large extent the standard account of the life of Ramanujan is a romantic invention and we consider our point made (see Kanigel [1991] for a more 'realistic' biography of Ramanujan.)

Therefore, if it is your ambition, as it is ours, to set the Kuhnian revolution in mathematics on its tracks, what to do (to quote a famous political philosopher)? It seems obvious to us that the first thing to do is to look for a good description of the subject itself: *what kind of thing is this curious process we call mathematical practice*? The aim of this book is two-fold:

- first, to bring together a number of authors who have thought and are still thinking about what mathematical practice is in general as well as in detail, how it should be studied and how theories can be formulated and,
- secondly, to incorporate existing materials from other, though related disciplines, as is, e.g., the case for mathematical education. This is a well-developed research community with its own goals, methods, and theories, but somehow it does not seem to connect all that well with the philosophical community. We wish to show in this book that such connections are indeed possible, if not necessary (if only for thought-economical reasons: duplication is rarely a time-energy saving device).

The papers presented here can thus be subdivided into three major categories:

- a first set deals with the general theme of mathematical practice,
- a second set with specific themes that arise when one takes the viewpoint from a full-blooded description of mathematical practice and
- a third set, too important to classify under the second heading and already referred to above, namely the relation between mathematical practice on research level, academic and otherwise, and education.

General theme: how to deal with mathematical practice?

In this section four papers have been brought together that show quite different ways of approaching the question above, showing thereby that the issue of how one is to study mathematical practice is itself a very difficult and complicated problem. Rather than looking for a unifying framework, it is our belief that by presenting four rather disparate approaches we will hopefully succeed in convincing the reader that it is not very likely that such a unifying theory will be easily put together, if at all.

Consider the paper by Jody Azzouni, "How and Why Mathematics is Unique as a Social Practice". Azzouni recognizes the importance of mathematical practice as a subject worthy of philosophical reflection and he tries to identify the characteristics that distinguish, quite sharply in his mind, mathematical practice from other practices. His inspiration and arguments are drawn from mainly analytical philosophy, more specifically from philosophers such as Ludwig Wittgenstein, Michael Resnik, Saul Kripke, and Hilary Putnam. Azzouni aims to show, so we believe, that the same analytical tools that were used to deny the importance of mathematical practice can also be used to make a case in its favour.

Eduard Glas in "Mathematics as Objective Knowledge and as Human Practice" attacks the problem of the nature of mathematical practice from a different angle, namely philosophy of science, more precisely the work of Karl R. Popper and to be even more precise, the three world model Sir Karl was so fond of. We quote from Glas' paper to show the Popperian mode of thinking present: "Humankind has used descriptive and argumentative language to create a body of objective knowledge, stored in libraries and handed down from generation to generation, which enables us to profit from the trials and errors of our ancestors." It allows Glas to reach the same goal as Azzouni, i.e., to show that mathematics can be properly distinguished from other practices and to see no deep conflict between mathematics' objectivity and its being profoundly social.

A similar concern is present in the contribution of Robert Thomas, "The Comparison of Mathematics with Narrative". However, Thomas walks a different route. Combining elements of semiotic theory – Umberto Eco, Brian Rotman, and Hayden White are referred to –, and of philosophy of mathematics and science – here we find Stephan Körner, Hartry Field and Hilary Putnam as sources of inspiration –, he too aims to show that, although mathematical practice shares a number of similarities to stories and narratives, nevertheless it is at the same time quite distinct.

Finally a deeply social and to many disturbing sound can be heard in Sal Restivo's "Theory Of Mind, Social Science, and Mathematical Practice". Starting from a theory of mind that in its 'classical setting' is typically asocial, Restivo shows how it impregnates our standard view of mathematics. Socialising the mind leads to a socialisation of mathematics. His main inspiration is to be found in the work of one of the founding fathers of sociology, viz. Emile Durkheim and his seminal notion of 'practices as institutions'. It leads him to the conclusion that a number of 'old' questions have to be posed again because entirely new answers are in the making. To quote from his paper: "What are numbers (and what are all the basic concepts and processes that constitute mathematics?). What is a classroom? What are teachers and students? What is learning? What is truth? What does it mean to reason? What is a proof? The trick here is to see all of these old friends as *institutions.*"

As said above, these contributions show that a genuine theory of mathematical practice is possible. Genuine in the sense that it is not derived from any foundational theory such as formalism, logicism or intuitionism (and all varieties of constructivism that it generated), to name the three major schools of the twentieth century. Such a theory is also badly needed and the four different ways presented here of handling the subject show at the same time that much work remains to be done in terms of mutual comparisons and enrichments. This present situation however need not be an obstacle in order to have a closer look at more specific elements of mathematical practice. Indeed, they help to refine the general problems and they provide detailed accounts that can be usefully employed as test cases for the diverse general accounts. That is the justification for the next major part of this book.

Specific themes: taking mathematical practice seriously

What kind of (more) detailed problems should one expect? What seems rather obvious is that the same problems that came up during the development of the philosophy of science in the post-Kuhn era, have a fairly large chance of appearing within the (new kind of) philosophy of mathematics as well. For that reason it was to be indeed expected that someone should write something about incommensurability. After all, incommensurability – the problem of whether or not it is possible to compare different scientific theories – has been a core problem in the philosophy of science, so at least one should have a look at it from the mathematical perspective. This is precisely what Otávio Bueno in his paper "Incommensurability in Mathematics" tries to do. Rather surprisingly perhaps, Bueno presents a strong case *in favour of* incommensurability in mathematics. Unlikely as it seems – after all, is not a number a number whenever and wherever it appears, so should not comparability be guaranteed at all places and at all times? – he does a wonderful job, presenting specific case studies to back up his claim. In his own words: "Theory change in mathematics, just as theory change in science, becomes a more complex, more interesting and not a cumulative phenomenon. As with science, in mathematics sensitivity to meaning change is required. This means that a simple cumulative pattern of mathematical development doesn't seem to make sense of mathematics." The unavoidable conclusion does follow: revolutions in mathematics are possible.

Who speaks of incommensurability, unavoidably has at the back of his or her mind the problem of (mathematical) progress. As soon as some form of incommensurability, however weak, sneaks in, the problem of how to define progress poses itself. Madeline Muntersbjorn in her contribution "Mathematical Progress as Increased Scope" tries to deal with this difficult question. The suggestion she proposes is that "…, mathematicians are not like mapmakers who adhere to the environmentalist's ethic, 'take only memories—leave only footprints.' They are more like *terraformers*, science-fiction engineers who travel to inhospitable planets and struggle to make alien landscapes suitable for human settlement by adapting them to our perceptual needs and abilities via innovations in formal systems of signification.". This to our minds powerful metaphor really begs to be further developed as it manages to steer a course between, on the one hand, the *Scylla* of strong forms of mathematical realism, including variations on Platonism, and, on the other hand, the *Charibdis* of relativism where "2 + 2 = ?" is a question on the same footing as "Did Sherlock Holmes have a homosexual affair with Watson?".

The next step, of course, must be to provide case studies from these new perspectives. Of course, if we are thinking about mathematics of the past, then the easiest thing to do is to look at the history and historians of mathematics. Surely they must have thousands of case studies ready for use. Although we happily accept – how could one argue otherwise? – the work done by historians, nevertheless we do have something slightly different in mind. What we are talking about is a shift of focus, looking at the neglected or almost forgotten details, connecting elements that seem unrelated at first sight, so that, if successful, the philosophical relevance becomes clear(er).

A fine example of such an attempt is Brendan Larvor's "Proof in C17 Algebra". Although he treats the well-known mathematicians of that period – Girolamo Cardano, François Viète, Thomas Harriot, John Pell, and others –, he does look at it in a different way. He is interested in what one could describe as "proof styles", making it possible to distinguish between a mathematical text of Viète in contrast with, say, Cardano. The connection with the philosophy of mathematics is easily made: what we count as proof today is something that took quite some time to develop and hence it is a delicate question to judge the *quality* of proofs. Hence, what are proofs? How can we distinguish proofs from arguments (if such a distinction is meaningful)? What will 'future' proofs look like (given that the proof concept is a mobile concept)?

Some of these questions, especially the question about proofs and arguments, are discussed in the contribution of Andrew Aberdein, "The Informal Logic of Mathematical Proof". It is perhaps a bit surprising to see the names of Stephen Toulmin and Douglas Walton appear in a paper about mathematical proofs. After all, are these two authors not famously known for their work in argumentation theory and definitely not in proof theory? And what could be the involvement of argumentation theory in the understanding of mathematical proof? Aberdein's claim is precisely that by looking at mathematics from an argumentation-theoretical point of view, aspects of the mathematical culture are brought into focus that, from a formal point of view, would be lost altogether. It thereby helps to refine our image(s) of mathematical practice. If proofs are situated elements of such practices, it would make much more sense to talk of *proof dialogues* as he proposes to do instead of proofs in an unqualified way, thereby inviting us to reify them.

There is however something extremely important that all contributions up to this point (mostly implicitly, sometimes explicitly) indicate: if mathematics is indeed a complex set of diverse mathematical practices, if indeed these practices are (also) shaped by social factors, dependent on societal circumstances, thus sensible to societal changes, and therefore very changeable, then to become a mathematician must also be constituted by a complex set of social processes. It cannot be a matter of 'simply' developing the faculties, capacities or powers already present in the genius' brain – "the seed is there, it merely needs to grow" –, rather it is a process that, to use a biological metaphor, aims at preparing an organism for a very specific environment. In short, any theory that takes itself seriously as a candidate for understanding mathematical practices, must deal with mathematics education. Hence the third part of this book.

The special case of mathematics education

As said above, the community of mathematics educators is a very well-established community of inquiry. However, it does not seem to connect very well with the philosophers of mathematics community (admitted that the latter group numberwise is rather small compared to the former one). Note that the intersection of the two sets is not empty: there are 'true' specialists, namely, philosophers of mathematics education. Nevertheless, it is our impression that they tend to be associated more strongly with the teachers than with the philosophers. In this third part of the book we wish to show that the two communities not only should, but really must meet more often and more intensely. In addition, it was our ambition to present the possible interactions between the philosophy of mathematics and mathematics education in as many ways as possible, from an abstract level to a very concrete level, from general considerations to case studies, from argumentation to narrative, from the institutional to the personal.

The first two papers of this part of the book are Leone Burton's "Mathematicians' narratives about mathematics – and their relationship to its learning" and Anthony and Dominic Peressini's "Philosophy of mathematics and mathematics education. The Confluence of Mathematics and Mathematical Activity". They both share the concern to show that mathematics and its philosophy on the one hand, and mathematics education and its philosophy on the other hand, have a lot to share. Do note that for both of them the idea to understand the proof concept as a social construct (we repeat, once again, without implying any form of deep relativism) is pivotal. For one thing, instead of the image whereby the ideal notion of logico-formal proof is transferred to the educational setting, we now have the image of a particular concept, viz. "proof", arising in a particular community, usually referred to as "the" mathematicians, and then being transferred into a totally different setting, namely, the teaching context. Seen thus, there is little need for an *exact copy* of the proof concept in the classroom. Hence, all kinds of questions pop up: What kind of proof is required for pupils to get a 'good' feeling for mathematics? How do other arguments (here Aberdein's paper is clearly relevant) function in the classroom? How do philosophical elements enter into the very same classroom?

It is more than obvious that answering these questions, apart from the philosophical setting, requires lots of case studies. We present here three such cases. We do know, of course, that there is wealth of materials available at the present moment. The emphasis, however, is not on a case study from the educational point of view, but from (at least) the philosophico-educational point of view. Jill Adler in her paper "Mathematical Practices in and across School Contexts" does precisely that. Analysing the situation in South Africa in the *post-apartheid* situation allows her to come to the conclusion that "... a decontextualised notion of mathematical practice makes no sense from the perspective of school mathematics, if at all. School mathematical practices are just that: practices dialectically produced by both mathematics and schooling."

The second example concerns the Belgian, more specifically Flemish situation. As both editors of this volume are working in Belgium, it seems quite normal to have a case study "close to home". However, as a case study, it is perhaps somewhat unusual and special because it involves a topic that is not often addressed, if at all, and equally often considered to be a borderline phenomenon. *In casu,* what we are talking about are journals for mathematics teachers, not to be confused with journals for educational mathematics, journals for philosophy of mathematics, journals for the philosophy of mathematics education, and so on. What is presented here in the paper by Ad Meskens, "The Importance of a Journal for Mathematics Teachers", are mathematics teachers writing for mathematics teachers. What do they write about? What do they consider to be so interesting that their colleagues should know about it? It is important to note that the author is a mathematician himself, not a philosopher. So here we have at least one example of a person-in-the-field reporting from the field.

The third and last example, also the concluding piece of this book (and the editors of this volume are truly proud to be able to include a contribution from this author) is basically about a formula and its proof, viz. Reuben Hersh's "On the Interdisciplinary Study of Mathematical Practice, with a Real Live Case Study". Formulated thus, this seems at first sight hardly innovative, refreshing or stimulating. However, in the unique style that is Hersh's own, we are invited to walk along with him and, indeed, think about a formula and its proof, but in such a way that at every step the links to education, to philosophy, and, of course, to mathematics become clear, and, in fact, refreshing and stimulating. It is worthwhile to state here his conclusion: "This much can be said. Mathematics really exists. It is going on, it is taking place, it has been around a long time and is here to stay. If your vocabulary insists that it is not real, and since in any ordinary meaning of the word it is not "fictional", then you must find some other kind of ontology, neither "real" in your sense nor fictional in any sense, to place it in."

In this sense, we have made a full circle, starting from philosophical considerations to the most concrete case study imaginable, back to philosophy with a refreshed mind.

A concluding remark and thought

First the remark: a special feature of this volume is that all the authors present here did actually meet physically at an international conference, *Perspectives on Mathematical Practices* (*PMP2002*), held at *Vrije Universiteit Brussel*, between 24 and 26 October 2002. It was organized by the *Centre for Logic and Philosophy of Science* (*CLWF*) at the same university (for the full program, see http://www.vub.ac.be/CLWF/PMP2002). This explains for a part the coherence among the diverse contributions, presented here. The other part explaining the coherence has to do with the fact that a selection has been made out of all contributions at the conference. However, the remaining contributions are not lost to the reader for these have been published in the logico-analytical journal *Logique et Analyse* (Van Kerkhove & Van Bendegem [2002]). The table of contents of this special volume can be found at: http://www.vub.ac.be/CLWF/L&A/. Finally, we like to mention that a discussion and mailing group has been launched at http://www.vub.ac.be/CLWF/mathprac/, to ensure that the debates will indeed continue.

And then the thought: there is a famous quote attributed to the famous French mathematician Jean Dieudonné (see [1982], p. 23): "Celui qui m'expliquera pourquoi le milieu social des petites cours allemandes du XVIIIe siècle où vivait Gauss devait inévitablement le conduire à s'occuper de la construction du polygone régulier à 17 côtes, eh bien, je lui donnerai une médaille en chocolat." ("The person who will explain to me why the social setting of the small German courts of the 18th century wherein Gauss lived forced him inevitably to occupy himself with the construction of a 17-sided regular polygon, well, him I will give a chocolate medal.") Although perhaps this book is not a straightforward answer to Dieudonné's worry, at least it seems reasonable to start to think about what kind of chocolate and what kind of medal we would like to have.

the editors,
Bart Van Kerkhove
Jean Paul Van Bendegem

REFERENCES

Jean DIEUDONNÉ: Mathématiques vides et mathématiques significatives. In: François Guénard & Gilbert Lelièvre (réd.), *Penser les mathématiques*. Paris: Editions du Seuil, 1982, pp. 15-38.

Robert KANIGEL: *The Man Who Knew Infinity. A Life of the Genius Ramanujan*. New York: Scribner's, 1991.

Bart VAN KERKHOVE & Jean Paul VAN BENDEGEM (guest editors): *Proceedings of the Perspectives on Mathematical Practices Conference 2002*. Special Issue of *Logique et Analyse*, vol. 45, nrs. 179-180, 2002.

I

HOW TO DEAL WITH MATHEMATICAL PRACTICE?

Chapter 1

HOW AND WHY MATHEMATICS IS UNIQUE AS A SOCIAL PRACTICE

Jody Azzouni
Tufts University, MA

Abstract: Difficulties are raised for views that explain consensus in mathematics using only sociological pressure. Mathematical proof is sociologically very peculiar, when compared to other socially constrained practices. A preliminary analysis of the factors that have been at work historically in the "benign fixation of mathematical practice" are then exhumed: dispositions, implicit applications, an implicit logic, all play a role.

Key words: Consensus, mathematical proof, socially-constructed objects, drift, mature mathematics, contemporary mathematics.

1.

I'm sympathetic to *many things* those who self-style themselves "mavericks" have to say about how mathematics is a *social* practice. I'll start with the uncontroversial point that mathematicians usually reassure themselves about their results by showing colleagues what they've done. But *many* activities are similarly (epistemically) social: politicians ratify commonly-held beliefs and behavior; so do religious cultists, bank tellers, empirical scientists, and prisoners.

Sociologists, typically, study methods of attaining *consensus* or *conformity*[1] since groups act in *concert*. And (after all) ironing out

[1] Attaining "conformity" and "consensus" are mild-sounding phrases for what's often a pretty *brutal* process. Although what I say is intended to be understood generally, the reader does best *not* to think of the practice of torturing political deviants (in order to bring them and their kind into line), but of doting parents teaching children to count, to hold forks, to maneuver about in clothing, or to speak.

B. Van Kerkhove and J.P. van Bendegem (eds.), Perspectives on Mathematical Practices, 3–23.

mathematical "mistakes" is suppressing a form of *deviant behavior*. One way to find genuine examples of socially-induced consensus is to limn the range of behaviors *possible* for such groups. One *empirically* studies, that is, how groups deviate from one another in their (group) practices. Consider admissible eating behavior. The options that *exist* are virtually *unimaginable*: in *what's* eaten, *how* it's eaten (in what order, with what *tools*, over how much time), how it's cooked—*if* it's cooked—what's allowed to be said (or not) during a meal, and so on. To understand why a group (at a time) eats meals as it does, and why its members find variants inappropriate (even *revolting*), we see how consensus is determined by childhood training, how ideology crushes variations by making them *unimaginable* or viscerally *repulsive* (so that, say, when someone imagines a *cheeseburger*, what's felt is—nearly instinctive—disgust), and so on. Equally coercive social factors in conjunction with the ones just mentioned explain why we obey laws, respect property (in the *particular* ways we do), and so on: The threat of punishment, corporeal or financial.

Before turning specifically to mathematical practice, note two presuppositions of any empirical study of the social induction of consensus (and which have been assumed in my sketchy delineation of the sociology of eating). First, such *social* induction presupposes (empirical) evidence of the *possibility* of alternative behaviors. The best way (although not the only way) to verify that a kind of behavior *is* possible is to find a group engaged in it; but, in any case, if a behavior is biologically or psychologically impossible, or if the resources available to a group prevents it, we don't need social restraints to explain why individuals *uniformly* avoid that behavior.

The second presupposition is that the study of social mechanisms should uncover factors powerful enough to exclude (in a given population) the alternatives we otherwise know are possible; either the absence of such factors, or the presence of empirical reasons that show such factors can't *enforce* behavioral consensus, will motivate the hypothesis of *internal factors*— psychological, physiological, or both—in conforming individuals: consider, for example, the Chomskian argument that internal *dispositions* in humans strongly constrain the general form of the rules for natural languages.

2.

Let's turn to mathematical practice. There are two striking ways it seems to differ from just about *any* other group practice humans engage in. One has been repeatedly noted by commentators on mathematics; the other, oddly, is (pretty much) overlooked.

It's widely observed that, unlike other cases of conformity, and where social factors *really are* the source of that conformity, one finds in mathematical practice *nothing like* the variability found in cuisine, clothing, or metaphysical doctrine. There *are* examples of deviant computational practices: Babylonian fractions or the one-two-many form of counting; but overall empirical evidence for the possible deviation from standard mathematical practice—at least until the twentieth century—isn't rich.

Two points. First, as Kripke and others have noted (in the wake of Wittgenstein),[2] it's easy to *design* thought experiments where people, impervious to correction, *systematically* follow rules differently from us. Despite the ease of *imagining* people like this, they're not *found* outside philosophical fiction. One *does* (unfortunately) meet people who can't grasp rules at all—but that's different. In rule-following thought experiments someone is portrayed who seems to follow *a* rule but who also understands "similar," so that she "goes on" differently from us. (After being shown a finite number of examples of sums, she sums new examples as we would until she reaches a particular border (pairs of numbers both over a hundred, say) whereupon she sums differently—in some systematic way—while claiming she's still doing the same thing. This really is different from people who don't grasp generalization at all.)

Despite the absence of *empirically real* examples of alternative rule-following (in counting or summing), such *thought experiments* are often used to press the view that it's (purely) *social factors* that induce mathematical consensus. Given my remarks about *appropriate* empirical methods for recognizing *real* options in group practices, such a claim—to be empirically respectable, anyway—can't batten on thought experiment alone; it needs an analysis of social factors that arise in *every* society that counts or adds—and which force humans to agree to the same numerical claims. The social factors that are pointed to, however, for example childhood learning, are ones shared by almost every other group practice (diet, language, cosmetics, and so on) which—in contrast to mathematical practice—show great deviation across groups. That is, even when systematic algorithmic rules (such as the ones of languages or games) govern a practice, that practice still drifts over time—unlike, as it seems, the algorithmic rules of mathematics.[3]

2 E.g.: Kripke 1982, Bloor 1983, and, of course, Wittgenstein 1953.

3 I should make this clear: by "drift" I mean a change in the rules and practices which doesn't merely involve augmentation of such rules, but the elimination of at least some of them. Mathematics is always being augmented; the point of denying "drift" in its case is that such augmentation is overwhelmingly monotonically increasing.

One possible explanation for this[4] is that practical exigencies exclude deviant rule-following mathematics: someone who doesn't add as we do can be exploited—in business transactions, say. And so it's thought that deviant counting would die quickly. But this idea is sociologically naive because, even if the dangerousness of a practice did imply its quick demise, this wouldn't mean it couldn't emerge to begin with, and leave evidence in our historical record of its temporary stay among us; all sorts of idiotic and quite dangerous practices (medical ones, cosmetic ones, practices motivated by religious superstition) are *widespread.* Even *shallow* historical reading exposes a plethora of, to speak frankly, pretty dumb activities that (i) allowed exploitation of all sorts (*and* helped shorten lives), and yet (ii) didn't require too much insight to *realize* were both pretty dumb and pretty dangerous. *There's* no shortage of such practices *today*—as the religious right and the raw-food movement, both in the United States, make clear. So it's hard to see why there can't have been really dumb counting practices that flourished (by, for example, exploiting the rich vein of number superstition we *know* existed), and then died out (along with the poor fools practicing them).

Another way around the apparent sociological uniqueness of mathematical practice is the blunt response that mathematical practice *isn't* unique; there *are* deviant mathematical practices; we just haven't looked in the right places for them. Consider, instead of *counting* variants, the development of alternative mathematics—intuitionism, for example, or mathematics based on alternative logics (e.g., paraconsistent logics). Aren't *these* examples of mathematical *deviancy* every bit as breathtakingly different as all the things people willingly put in their mouths (and claim tastes good and is good for them)?

Well, *no*. What should strike you about "alternative mathematics"—unless you're blinded by an a priori style of foundationalism, where a specific style of mathematical proof (and logic), and a specific subject matter, are definitional *of* mathematics—is that such mathematics is *mathematics as usual*. One mark of the ordinariness of the stuff is that contemporary mathematicians shift in what they prove results about: they practice one or another branch of "classical" mathematics, and then try something more exotic—if the mood strikes. Proof, informal or formal, looks like the same thing (despite principles of proof being *severely* augmented or diminished in such approaches).

[4] See e.g., Hersh 1997, p. 203.

Schisms among mathematicians, prior to the late nineteenth century, prove even *shallower* than this.[5] That differences in methodology historically prove divisive can't be denied: differences in the methodology of the calculus, in England and on the continent, for example, *retarded* mathematical developments in England for over a century. Nevertheless, one finds British mathematicians (eventually) adopting the continental approach to the calculus, and doing so because they (eventually) recognized that the results they wanted, and more generally, the development of the mathematics surrounding the calculus, were easier given continental approaches. British mathematicians didn't deny the cogency of such results on the grounds that the methods that yielded them occurred in a "different (incomprehensible) tradition."

Let's turn to the second (*unnoticed*) way that mathematics *shockingly* differs from other group practices. *Mistakes are ubiquitous in mathematics.* I'm not *just* speaking of the mistakes of professional—even brilliant—mathematicians although, notoriously, they make *many* mistakes;[6] I'm speaking of *ordinary* people: they find mathematics *hard*—harder, in fact, than just about any other intellectual activity they attempt. What makes mathematics difficult is (1) that it's *so easy* to blunder in; and (2) that it's *so easy* for others (or oneself) to see—when they're pointed out—that blunders *have* been made.[7]

So? This is where it gets cute. When the factors forcing behavioral consensus are genuinely social, *mistakes can lead to new practices.* This is for two reasons at least: first, because the social factors imposing consensus are often blind to details about the behaviors enforced—they're better at imposing uniformity of behavior than at pinning down *exactly which* uniform behavior the population is to conform to. If enough people make a certain mistake, and if enough of them pass the mistake on, the social factors enforcing consensus continue doing so despite the shift in content. Social mechanisms that impose conformity are good at synchronic enforcement; they're not as good at diachronic enforcement. (Thus what's sometimes described as a "generation gap.")[8]

5 One point of this paper is to provide an explanation for why this should be so; see what's forthcoming, especially section 6.

6 This is especially stressed in the "maverick tradition" to repeatedly hammer home the point that proof doesn't confer "certainty."

7 What makes mathematics hard is both how easy it is to make mistakes and how difficult it is to hide them. Contrast this with poetry. It's as easy to make mistakes in poetry—write stunningly bad poetry—as it is to blunder in mathematics. But it's much easier to cover up poetic blunders. Why that is is extremely interesting, but something I can't fully get into now.

8 Consider school uniforms. All sorts of contingent accidents cause mutations in such uniforms; but that (at a time) the uniforms should be, um, uniform, is a requirement. It's

The second reason is that the power of social factors to enforce conformity often turns on the successful *psychological internalization* of social standards; but if such standards are imperfectly internalized (and *any* standard—however mechanical, i.e., algorithmic—can be imperfectly internalized), then the *social standards themselves* can evolve, since, in certain cases, nothing else fixes them. Two examples are, first, the drift in natural languages over time: this is often because of systematic mishearings by speakers, or interference phenomena (among internalized linguistic rules), so that certain locutions or sounds drop out (or arise). The second example is when an external standard supplementing psychological internalization of social standards is operative, and is *taken to* prevent drift—for example holy books. Notoriously, such things are open to *hermeneutical drift*: the subject population reinterprets them (often inadvertantly) because of changes in language, "common sense," and therefore changes in their (collective) view of what a given law-maker (e.g., God) *obviously* had in mind.

In short, although every social practice is easy to blunder in, it's not at all easy to get people to recognize or accept *that* they've made mistakes (and therefore, if enough of them do so, it's impossible—nearly enough—to *restore* the practice as opposed to—often inadvertantly—starting a new one).

The foregoing focus on mistakes *isn't* meant to imply that *conscious* attempts to change traditions aren't effective: of course they can be (and often are). But mathematical practice resists *willful* (deliberate) change too. A dramatic case of a conscious attempt to change mathematical practice which failed (in large part because of incompetence at the standard fare) is Hobbes[9]. Another informative failure is Brouwer, because Brouwer was *anything but* incompetent at the standard practice.

Notice the point: Brouwer wasn't interested in developing *more* mathematics, nor were (and are) the other kinds of constructivists that followed; he wanted to change the *practice*, including his own earlier practice. But he only succeeded in developing *more* mathematics, not in changing *that* practice (as a whole). This makes Hilbert's response to Brouwer's challenge, by the way, *misguided*, because Hilbert's response was also predicated on (the fear of) Brouwer inducing a change in the practice. This is common: fads in mathematics often arise because someone (or a group, e.g., Bourbaki) thinks that some approach can become *the* tradition of mathematics—the result, invariably, is just *more* (additional) mathematics. A related (sociological) phenomenon is the mathematical *kook*—there

very common for a population to slowly evolve its culinary practices, dress, accent, religious beliefs, etc., without realizing that it's doing so.

9 See Jesseph 1999.

enough of these to write *books* about.[10] Only a field in which the recognition of mistakes is extremely robust can (sociologically speaking) successfully marginalize so many otherwise competent people *without* standard social forms of coercion, e.g., prison.

So (to recap.) mistakes in mathematics are common, and yet mathematical culture doesn't splinter because of them, or *for any other reason* (for that matter)[11]; that is, permanent *competing* practices don't arise as they can with other socially-constrained practices. This makes mathematics (sociologically speaking) *very odd.* Mathematical standards—here's another way to put the point—are robust. Mistakes *do* persevere, of course; but mostly they're eliminated, even when *repeatedly* made. More importantly, mathematical practice is so robust that even if a mistake eludes detection for years, and even if many results are built on that mistake, this *won't* provide enough social inertia—once the error *is* unearthed—to resist changing the practice back to what it was originally: in mathematics, even after lots of time, the subsequent mathematics built on the "falsehood" is repudiated.[12]

This aspect of mathematical practice has been (pretty much) unnoticed, or rather, *misdescribed*; and it's easy to see how. If one focuses on *other* epistemic issues, scepticism say, one can confuse the rigidity of group standards in mathematics with the availability of *certainty*: one can claim that, if only one is *sufficiently careful*, *really* attends to each step in a proof, carefully analyzes proofs so that each step immediately follows from earlier

[10] For example, Dudley 1987.

[11] Philip Kitcher, during the discussion period on November 21, 2002, urged otherwise—not with respect to mistakes, but with respect to conscious disagreement on method: he invoked historical cases where mathematicians found themselves arrayed oppositely with respect to methodology—and the suggestion is that this led to schisms which lasted as long (comparatively speaking) as those found among, say, various sorts of religious believers: one thinks (again) of the controversy over the calculus, or the disputes over Cantor's work in the late nineteenth century. What's striking—when the dust settles, and historians look over the episodes—is how nicely a distinction may be drawn between a dispute in terms of proof procedures and one in terms of admissible concepts. The latter sort of dispute allows a (subsequent) consistent pooling of the results from the so-called disparate traditions; the former does not. Thus there is a sharp distinction between the (eventual) outcome of disputes over the calculus, and (some of) those over Cantor's work. The latter eventually flowered into a dispute over proof procedures which proved irresolvable in one sense (the results cannot be pooled) but not in another. See 6.

[12] Contrast this with our referential practices: Evans (1973, p. 11) mentions that (a corrupt form of) the term "Madagascar," applied to the African mainland, was mistakenly taken to apply to an island (indeed, the island we currently use the term to refer to). Our discovery of this error doesn't affect our current use of the term "Madagascar"—the social inertia of our current referential practice trumps any social mechanisms for correcting dated mistakes in that practice.

ones, dutifully surveys the whole repeatedly until it can be intuited in a flash, then one can rig it so that—in mathematics, at least—one won't *ever* make any mistakes *to begin with*: one can be totally *certain*.'[13]

But there's a number of, er, mistakes in this Cartesian line. First, it's a robust part of mathematical practice that mistakes are found and corrected. Even though the practice is therefore *fixed* enough to rule out deviant practices that would otherwise result from allowing such "mistakes" to change that practice, this *won't* imply that *psychologically-based* certainty is within reach. For it's compatible with the robustness of our (collective) capacity to correct mathematical mistakes that some mistakes are still undetected—even old ones.

Apart from this, the psychological picture the Cartesian recipe for certainty presupposes is inaccurate. It's very hard to correct your own mistakes, as you know, having proofread your work in the past. *And* yet, someone else often sees *your* mistakes at a glance. This shows that the Cartesian project of gaining certainty *all alone*, a strategy crucial for Descartes' demon-driven epistemic program, is quixotic.[14]

Notice, however, that the Cartesian view would explain, *if it were only true*, how individuals can disagree on an answer, look over each other's work, and then come to agree *on what the error is*. (They become CERTAIN of THE TRUTH, and THE TRUTH is, after all, THE SAME.) Without this story, we need to know what practitioners have internalized (psychologically) to allow such an unnaturally agreeable social practice to arise.[15]

[13] And then one can make this an epistemic requirement on all knowledge (and offer recipes on how to carry it off). Entire philosophical traditions start this way.

[14] One of the ways Newton is so remarkable is that he did so much totally on his own, by obsessively going over his own work. (See Westfall 1980.) Newton's work is an impressive example of what heroic individualistic epistemic practice can sometimes look like. Despite this, Newton made mistakes.

[15] Unlike politics, for example, or any of the other numerous group activities we might consider, mathematical agreement isn't coerced. Individuals can see who's wrong; at least, if someone is stubborn, others (pretty much all the competent others) see it. Again, see Jesseph 1999 for the Hobbesian example. Also recall Leibniz's fond hope that this genial aspect of mathematical practice could be grafted onto other discourses, if we learned to "calculate together." By contrast, Protestantism, with all its numerous sects—in the United States especially—is what results when coercion isn't possible (because deviants can, say, move to Rhode Island). And much of the history of the Byzantine empire with its unpleasant treatment of "heretics" is the normal course of events when there's no Rhode Island to escape to. It's sociologically very surprising that conformity in mathematics isn't achieved as in these group practices. Imagine—here's a dark Wittgensteinian fable—we tortured numerical deviants to force them to add as we do. (Recall, for that matter, George Orwell's 1984.)

To summarize: What seems odd about mathematics as a social practice is the presence of substantial conformity on the one hand, and yet, on the other, the absence of (sometimes brutal) social tools to induce conformity that routinely appear among us *whenever* behavior really is socially constrained. Let's call this "the benign fixation of mathematical practice."

3.

The benign fixation of mathematical practice *requires* an explanation. And (it should be said) Platonism is an appealing one: mathematical objects have their properties necessarily, and we perceive these properties (somehow). Keeping our (inner) eye firmly on mathematical objects keeps mathematical practice robust (enables us to find mistakes). The problem with this view—as the literature makes clear—is that we can't explain our epistemic access to the objects so posited.[16]

One might try to finesse things: demote Platonic objects to socially-constructed items (draw analogies between numbers and laws, language, banks, or Sherlock Holmes). Address the worry that socially-constructed nonmathematical objects like languages, Mickey Mouse, or laws, *evolve over time* (and that their properties look conventional or arbitrary) by invoking the content of mathematics (mathematical rules have content; linguistic rules are only a "semitransparent transmission medium" without content). And, claim that such content makes mathematical rules "necessary."[17]

As this stands, it won't work: we can't bless necessity upon whatever we'd like by chanting "content." Terms that refer to fictional objects have content too—that doesn't stop the properties attributed to such "things" from *evolving* over time; socially-constructed objects are *our* objects—if we take their properties to be fixed, that's something we've (collectively) imposed on them. It's a good question *why* we did this with mathematical terms, and not with other sorts of terms.

If socially-constructed objects are stiffened into "logical constructions" of some *fixed* logic plus set theory (say), this doesn't solve the problem: one

16 Current metaphysics robs Platonism of respectability. Sprinkle mysticism among your beliefs, and the perceptual analogy looks better; introduce deities to imprint true mathematical principles in our minds, and the approach also looks appealing. Deny all this, and Platonism looks bizarre.

17 See Hersh 1997, p. 206. I deny that (certain) socially-constructed objects, mathematical objects and fictional objects, in particular, exist in any sense at all. See my 2004a. Nominalism, though, won't absolve me of the need to explain the benign fixation of mathematical practice. On the contrary.

still must explain how logic (of whatever sort) and set theory accrue social rigidity (why won't we let our set theory and logic *change*?).

There is no simple explanation for the benign fixation of mathematical practice because, as with any group practice, even if that practice retains its properties over time, that doesn't show that it has those properties (at different times) for the same reasons. Mathematical practice, despite its venerable association with *unchanging objects*, is a historical entity with a long pedigree, and so the reasons for why the correction of mistakes, for example, is robust in early mathematics, are *not* the same reasons for that robustness *now*.

4.

So now I'll discuss a number of factors, social and otherwise, and speculate how (and when) they contributed to the benign fixation of mathematical practice. The result, interestingly, is that if I'm right, benign fixation is historically contingent (and complex). That's a surprise, I suppose, for apriorists, but not for those of us who long ago thought of mathematics as something *humans* do over time.

Let's start (as it were) at the beginning: the historical emergence of mathematical practice (primarily counting and sums), and as that practice appears today among people with little or no other mathematics. Here it's appropriate to consider the role of "hard-wired" psychological dispositions. There seem two such relevant kinds of disposition. The first is a capacity to carry out algorithms, and—it's important to stress—this is a species-wide capacity: we can carry out algorithms, and teach each other to carry out (specific) algorithms *in the same way*. That's why we can play games *with each other* (as opposed to *past* each other), and why we can teach each other games that we play alone in the same way (e.g., one or another version of solitaire).

I can't say what it is about us—neurophysiologically I mean—that enables us to carry out algorithms the same way—no one can (yet); it's clear that some of us are better at some algorithms than others (think of games, and how our abilities to play them varies)—but what's striking is that those of us who are better aren't, by virtue of *that*, in *any* danger of being regarded as *doing something else*.

In describing us as able to "carry out" the same algorithms, I don't mean to say that we're *executing* the same algorithms (otherwise our abilities to carry out algorithms wouldn't *differ*). I understand "executing an algorithm," as doing (roughly) what a Turing machine does when it operates. Perhaps humans do something like that with *some* algorithm(s) or other (but, surely, different humans execute different algorithms). In any case, when we learn arithmetic, for example, we're *not* learning to *execute* any of the numerous (but equivalent) algorithms that officially characterize arithmetic operations—instead what we're learning is what a particular algorithm *is*, and how to *imitate* its result—or at least *some* of its results—by actions of our own. So when I describe us as "carrying out" an algorithm **A**, I mean that we're imitating it by doing something else **B**, not by executing *it*.

I mean this. Add two numbers fifteen times, and you do *something different* each time—you do fifteen *different* things that (if you don't blunder) are the same *in the respect needed*: the sum you write down at the end of each process is the same (right) one. *We* can't do *anything* twice; it's only, as it were, *parts* of our behavior (at a time) that occur repeatedly. Proof? We *remember* much of what we do, and we're never but never *just* imitating a numerical algorithm when we do so; we're squirming in our chairs, taking in some of the passing scene (through our *ears*, if not through our eyes), etc. *Machines* execute algorithms and can do so by doing some things *twice*. We're (I hope this isn't news) *animals*.

Another way to make the same point about how we imitate rather than execute the algorithms that we're officially working on is that our learning such algorithms enjoys an interesting flexibility: we not only (apparently) acquire and learn new algorithms, but we can get better at the algorithms we've already learned by practicing them.

Finally, it should be noted that, usually, mathematicians don't execute the algorithms they're officially deriving results from; they *short-cut* them.[18]

These considerations—phenomenological ones, it's true—suggest that we don't—probably can't—execute the official algorithms we're carrying out; we're executing other algorithms instead that imitate the target algorithm (and over time, no doubt, different ones are used to do this); and this neatly explains why we can improve our abilities, by practice, to add sums, carry out other mathematical algorithms, and win games (for that matter).

Having said this, I must stress that I'm speculating about something that must ultimately be established empirically. So (of course) it could turn out that I'm just *wrong*, that we really do execute (some of) these algorithms (or, at least, some subpersonal part of us does), and that we don't imitate them via other algorithms that we execute. The neurophysiologists, in the end, will tell us what's what (if it's possible for *anyone* to tell us this, I mean): I'm

[18] Our ability to imitate algorithms flourishes into mathematical genius (in some individuals, anyway); for the mathematician, as I said, never (or almost never) figures out what an algorithm (proof procedure, say) will yield by executing that algorithm directly. Ordinary mathematical proof—its form, I mean—already shows this. See my 2004b for more details.

betting, however, on my story—it explains our algorithmic flexibility, and our capacity to make and correct mistakes, in a way that a story that requires us to actually execute such algorithms doesn't.[19]

I'm also unable to say—because this too is ultimately a matter of neurophysiology—how *general* our capacity to mimic algorithms is; that we can now (since Turing and others) formulate in full generality the notion of mechanized practice—algorithm—doesn't mean that we have the *innate* capacity to "carry out"—imitate the results of *any* such algorithm whatsoever. Our capacity to imitate algorithms may be, contrary to (introspective) appearance, more restricted than we realize.

A (species-wide) capacity to imitate the execution of algorithms *in the same way* doesn't explain the benign fixation of mathematical practice. This is because that robustness turns on *conserving* the official rules governing mathematical objects, and a group ability to imitate algorithms the same way won't explain why a practice *doesn't* evolve by *changing* the applicable algorithms altogether—in just the way that languages, which involve algorithms too, evolves.

A second innate capacity I'm willing to attribute to us is a disposition to *execute* certain *specific* algorithms.[20] I'm still thinking here primarily of our (primitive) ability to count and handle small sums. My suggestion is that why, wherever primitive numerical practices emerge, they're (pretty much) the same isn't because of sociological factors that constrain psychologically possible variants—rather, it's because of fixed *innate dispositions*.

Don't read too much into this second set of dispositions since they're also too weak to explain the benign fixation of mathematical practice: they don't extend far enough. They're *not* rules that apply to, say, any counting number whatsoever. These dispositions—I suspect—are very specific: they may facilitate handling certain small sums by visualizing them, or manipulating tokens in certain ways. I'm *not* claiming that such dispositions enable the

[19] The last four paragraphs respond to a line of questioning raised by David Albert; my thanks for this.

[20] Caveat: Given my earlier remarks about the empirical nature of my speculations about how we imitate the execution of algorithms, I'm not sure I've succeeded in describing two distinct capacities: Our ability to imitate algorithms—in general—needn't be a general ability to execute algorithms because, as I've said, we don't "carry out" algorithms by directly executing them. What we do, perhaps, is apply a quite specific algorithm or set of algorithms to official algorithms that we want to carry out, a process which enables us to extract (some) information about any algorithm (once we've psychologically couched it a certain way).

execution of (certain) algorithms so that we can count as high as we like, add arbitrarily large sums, and so on.[21]

5.

Such innate dispositions as I've described, although they explain why the independent emergence of counting and summing among various populations always turns out the same, won't explain why, when mathematics becomes professionalized—in particular, when *informal deduction* is hit upon—benign fixation continues, rather than mathematical practice splintering.

I introduce something of a sociological idealization, which I'll call "mature mathematics," and which I'll describe as emerging somewhat before Euclid and continuing until the beginning of the twentieth century.[22] I claim that several factors conspire to benignly fix mature mathematical practice.

The first is that, *pretty much until the twentieth century*, mathematics came with intended *empirical domains of application* (from which mathematical concepts so applied largely arose). Arithmetic and geometry, in particular, come with *obviously* intended domains of application. These fixed domains of application prevent, to some extent, drift in the rules governing terms of mathematics—in these subjects so applied, anyway. This is because *successful application* makes us loathe to change successfully applied theorems—if that costs us applicability.[23]

But something more must be going on with *mathematics*, as a comparison with empirical science indicates. For the history of empirical

[21] Thus I haven't (entirely) deflected Kripkean attacks on the dispositionalist approach to rule-following. And: On my view, dispositions have only a partial role in the benign fixation of mathematical practice.

[22] Twentieth-century mathematics isn't mature? Well, of course it is, but I'm arguing that it's different in important respects that require distinguishing it (sociologically, anyway) from what I've called "mature mathematics." Maybe—taking a nomenclatural tip from literature studies—we can call it "post-mature mathematics." On the other hand, maybe we better not. I'll contrast "mature" mathematics with "contemporary" mathematics. This idealization is artificial because aspects of mathematical practice present in "mature" mathematics (they're in Euclid), and which continue to play an important role in contemporary mathematics, don't fit my official characterization of mature mathematics. I'll touch on this in due course.

[23] The ancient Greeks, it's pointed out more than once, were disdainful of "applied mathematics." Yes, but that disdain is compatible with what I've just written. The view, for example, that the empirical realm is a copy of the mathematical realm both determines the intended empirical domain in the sense I mean, while simultaneously demeaning the intellectual significance of that domain.

science (physics, in particular) proves that drift *can* occur and yet the intended application of the concepts and theories not vanish as a result. Newtonian motion, *strictly speaking*, occurs only when objects *don't move*. But its *approximate* correctness suffices for successful application. Furthermore, the application of mathematics—geometry especially—always involves (some) approximation because of the nature of what geometric concepts are applied to (fuzzily-drawn figures).

Fixing mathematical practice beyond the drift allowed by successful (but approximate) application, is a crucial factor, viz., the essential role of *informal proof*, or deduction. It's no doubt debatable exactly what's involved in informal proof, but in mature mathematics it can be safely described as this:[24] a canonization of logical principles, and an (open-ended) set of additional (mathematical) principles and concepts which, (1) (partially) characterize subject matters with intended domains of application, (2) are more or less tractable insofar as we can, by means of them, informally prove new unanticipated results, and (3) which grow monotonically over time.

The need for tractable informal proofs drives the existential commitments of mathematics, and in particular, drives such commitments *away from* the objects characterized (empirically) in the intended domains of application. I've described this process in two case studies elsewhere, and won't dwell on it now.[25] But the particular form mathematical posits take itself now contributes several ways to benign fixation.

First, ordinary folk practices with empirical concepts allow those concepts to drift in what we can claim about them, and what they refer to, without our taking ourselves, as a result, to be referring to something new. If we discover that gold is actually blue, we describe that discovery in exactly those words (and not as a discovery that there is *no* gold).[26] By taking mathematical posits as empirically uninstantiated items, we detach mathematical language from this significant source of drift in what we take to be true of them.

Second, once mathematical posits are taken to be *real* but sensorily unavailable items which provide truths successfully applicable to empirical domains, mathematical practice opens itself to philosophical concerns both about the nature of such truths and how such truths *are established*. For a

[24] Recall, however, footnote 22.

[25] See my 2004a and my 2004b. A discussion of the special qualities that a set of concepts, and principles governing them, needs for amenability to mathematical development, qualities that empirically derived notions, and truths about them, usually don't have, may be found in my 2000a.

[26] I'm alluding here to the sorts of thought experiments Putnam gave in his 1975. See my 2000b, especially Parts III and IV. There are subtleties and complications with this view of empirical terms, of course; but they don't affect points made in this paper.

number of reasons—mostly involving various philosophical prejudices about *truth*[27]—the conventionalist view that mathematical truths are stipulated, and that mathematical objects exist in no sense at all, isn't seen as tenable (or even *considered*), and a view of eternal and unchanging mathematical objects carries the day instead.[28]

Of course, such an eternalist view of mathematical objects doesn't, all on its own, eliminate the possibility of a mathematical practice which allows drift in what we take to be true of mathematical objects: we could (in principle) still allow ourselves to be wrong about mathematical objects, and to be willing to change basic axioms governing them as a result. Imagine this thought experiment: after we discover nonEuclidean geometry, we decide that Euclidean geometry is wrong; that is, we take ourselves to have been wrong about geometric *abstracta*—there are no abstracta that obey Euclidean axioms. This attitude is compatible with a view of mathematical abstracta as eternal, unchanging, etc. What prevented it from emerging, I claim, is the (historically speaking) relatively *late* discovery of nonEuclidean geometry. I touch on this later, but my view is that had (one or another) nonEuclidean geometry emerged in, say, ancient times, and had it been the case that Euclidean geometry proved useless in its intended domain of application (in comparison to nonEuclidean geometry), it would have been supplanted by nonEuclidean geometry—we would have taken ourselves to have been wrong about geometry and would have changed the basic axioms of what we called *Geometry* to suit.[29]

[27] I have in mind claims like: (i) we can't have truths about things that don't exist, and (ii) even if we could, such truths wouldn't prove as empirically useful as mathematics is. Such prejudices are hardly restricted to ancient philosophers—e.g., Plato and Aristotle. They are standard fare among contemporaries too. See my 2004a for what things look like once we purge ourselves of them.

[28] Philosophical views about what positions are sensible or not can't be ignored in any sociological analysis of why a group practice develops as it does. There are some, no doubt, who take philosophical views as mere ideology, as advertising for other more substantial social motives (e.g., professional or class interests). I can't see how to take such a position seriously, especially if it's the sociology of knowledge-practices (of one sort or another) that's under study. What looked philosophically respectable, or not, I claim, had (and has) a profound impact on mathematical practice. It may be a mistake to search for that effect in the theorem-proving practices of the ordinary mathematician, but, in any case, as this paper illustrates, I locate it in, as it might be described, the general framework of how mathematics operates as a subject-matter (in particular, in how it's allowed to change over time).

[29] This would have happened, in part, because of an implicit metaphysical role for mathematical objects in the explanation for why that mathematics applied to its intended empirical domain—recall the resemblence doctrine mentioned in footnote 23. But in part I attribute the late emergence of nonEuclidean geometry not having a supplanting effect on Euclidean geometry as due to the already in place change of "mature" mathematics into

I've stressed how intended domains of application helped to benignly fix mathematical practice; the (implicit) canonization of logical principles is just as essential. Had there been shifts in the (implicit) logic, then we would have found ourselves—when considering early mathematics—in exactly the same position that modern Greeks find themselves if they try to read ancient Greek on the basis of their knowledge of the contemporary stuff: incomprehension. In addition, shifts in the implicit logical principles utilized would have led to incompatible branchings in mathematical practice because of (irresolvable) disagreements about the implications of axioms and the validity of proofs.

6.

In order to motivate my discussion of how twentieth-century mathematics differs from what came before, I need to amplify my claim about the (tacit) canonization of logical principles in mature mathematics. Contemporary discussions of Frege's logicist program, and the *Principia* program that followed it, often dwell—quite melodramatically—on paradox; and the maverick animus towards such projects focuses on the set-theoretic foundationalism that's taken to have undergirded both the ontological concerns and the obsession with rigor proponents of such programs expressed.[30] But this focus obscures what those projects *really showed*: Nothing about the (real) subject-matter of mathematics (I rush to say), for *that's* proven to be elusive in any case—ways of embedding systems of mathematical posits in other systems is so unconstrained, ontologically speaking, that it's inspired structuralist views of that ontology.[31]

However, a very good case can be made that the *logic* of mature mathematics *was* something (more or less) equivalent to the first-order predicate logic, and that this was a nontrivial thing to have shown.[32] What

"contemporary" mathematics, as I characterize the latter shortly. I guess I'm hypothesizing a "paradigm shift" although I don't much like this kind of talk. It seems that Kline (1980) is sensitive to some of the changes from mature to contemporary mathematics, although he takes a rather darker view of the shift than I do.

30 In describing the complex history this way, I'm not necessarily agreeing either with the depiction or with the attribution of these motives to later proponents of set-theoretical foundationalism.

31 See, e.g., Resnik 1997.

32 Why fix on first-order logic, and not a higher-order (classical) logic, especially since it was a higher-order logic that historically arose first? Well, there are a number of reasons; but the one pertinent to the topic of this paper is that first-order logic and higher-order logics are nicely distinguishable because first-order logic is a canonization of reasoning

proves this is that the project of characterizing (classical) mathematics axiomatically in first-order classical systems succeeded.[33] What shows it isn't a trivial point is that, in fact, much of twentieth-century mathematics *can't* be so axiomatically characterized.

What's *especially* striking about this success is that the classical logic which is the algorithmic skeleton behind informal proof remained tacit until its (late) nineteenth-century *uncovery* (I coin this word deliberately). But, as the study of ever-changing linguistic rules shows, implicit rules have a slippery way of mutating; in particular, what seems to be a general rule (at a time) can subsequently divide into a set of *domain-specific* rules, only *some* of which are retained.[34] The logical principles implicit in mathematical practice—until the twentieth century, however—remained the *same* topic-neutral ones (at least relative to mathematical subject-matters). Such uniformity of logical practice suggests, as does the uniformity of counting and summation practices I discussed earlier, a "hard-wired" disposition to reason in a particular way.[35]

This brings us to the points I want to make about twentieth- century mathematics. *Contemporary mathematics*, I claim, breaks away from the earlier practice in two extremely dramatic respects. First, it substitutes for classical logic (the tacit canon of logical principles operative in "mature" mathematics), proof procedures of *any* sort (of logic) *whatsoever* provided only that they admit of the (in principle) mechanical *recognition* of completely explicit proofs. That is, not only are alternative logics, and the

principles without a subject matter, and the higher-order stuff (pertinent to this discussion) isn't: such "logics" are both open-ended, and (implicitly) saddled with a subject-matter.

33 Hersh (1997) and other mavericks deny this but offer only the (weak) argument that the project hasn't been carried out in detail for all the mathematics it was supposed to apply to. But why is that needed? (The same grounds show, I suppose, that Gödel's second incompleteness theorem hasn't been shown either; and there are other examples in mathematics as well.) By the way, notice that it's irrelevant that the ordinary mathematician neither now, nor historically, couched any of his or her reasoning in such a formalism. This is because—as mentioned earlier—nobody carries out an algorithm by executing that algorithm—especially not gifted mathematicians who strategize proofs (and their descriptions) routinely, if they give proofs at all.

34 See, e.g., Anderson 1988, especially pp. 334-335.

35 There is a complication that (potentially) mars this otherwise appealing view of the implicit role of first-order logic in mathematics: the "logic" of ordinary language looks much richer than what the first-order predicate calculus can handle—notoriously, projects of canonizing the logic of anything other than mathematics using (even enrichments of) the first-order predicate calculus have proved stunningly unsuccessful. This leaves us without a similar argument that the tacit logic of natural language is (something similar to) the first-order predicate calculus. But it would be very surprising if the tacit logic of mathematics were different from that of ordinary language—especially given the apparent topic-neutrality of that logic. I can't get further into this very puzzling issue now.

mathematics based on such things, *now* part of contemporary mathematics; but various sorts of *diagrammatic* proof procedures are part of it as well; such proof procedures, which involve conventionalized moves in the construction of diagrams, need not be proofs easily replicated in language-based axiomatic systems of any sort.[36]

One factor that accelerated the generalization of mathematical practice beyond the tacit classical logic employed up until the twentieth century was the explicit formalization of that very logic. For once (a version of) the logic in use was made explicit, mathematicians could *change it*. Why? Because what's *conceptually central* to the notion of *formal proof*, and had been all along (as it had been operating in mature mathematics), isn't the presence of any particular *logic*, or logical axioms of some sort, but only the idea that such a thing is *mechanically recognizable*.[37] This is neat: since (until the late nineteenth century) the logic *was* tacit, its particular principles couldn't have been seen as essential to mathematical proof since they weren't seen *at all*. What *was* seen clearly by mathematicians and fellow-travelers (recall footnote 15) was the benign fixation of mathematical practice; but *that's* preserved by generalizing proofs to anything algorithmically recognizable, *regardless* of the logic used.[38]

[36] There's lots more to say about diagrammatic proofs, but not now. I've now discharged the promissory note of the second paragraph of footnote 22, however: Diagrammatic proofs are in Euclid's elements (see my 2004b), and they continued to appear in mathematics during its entire mature phase—even though practices using them are only awkwardly canonized in a language-based theorem-proving picture of mathematical practice. The discussion of such items in contemporary mathematics is showing up in the literature on mathematical method. See, e.g., Brown 1999. I should stress again, however, that such practices require mechanical recognition of proofs; so they nicely fit within my (1994) characterization of mathematics as an interlocking system of algorithmic systems. I should also add that diagrammatic practices within classical mathematics are clearly compatible with the tacit standard logic used there—they provide consistent extensions of the axiomatic systems they accompany (or so I conjecture)—something not true of the more exotic items (e.g., logics) invented in the twentieth century.

[37] Informal proofs are sketches that algorithmically recognizable proofs exist, so the former don't presuppose a particular logic either. Notice that it's certainly not a requirement on an informal proof that it can be replaced by a (surveyable) formal proof. But why should this be a requirement? See my 2004c for more of a defense of this position.

[38] Haim Gaifman (November 21, 2002) has raised a challenge to the idea that contemporary mathematics can (genuinely) substitute algorithmic recognizability for the implicit logic of mature mathematics. For given the fact—aired previously—that mathematicians don't execute the actual algorithmic systems they prove results from, it must be that they rely on methods (modeling, adopting a metalanguage vantage point, etc.) which incorporate, or are likely to, the classical logic mathematicians naturally (implicitly) rely on. This suggests that if a mathematician were to attempt to really desert the classical context (and not merely avoid a principle or two—as intuitionists do), he or she would have to execute such algorithms mechanically—any other option would endanger the validity of the results

The second way that contemporary mathematics bursts out from the previous practice is that it allows pure mathematics *such* a substantial life of its own that areas of mathematics can be explored and practiced without even a *hope* (as far as we can tell) of empirical application.[39] This, coupled with the generalization of mathematical proof to mechanical recognition procedures (of one sort or another) allows a *different* way to benignly fix mathematical practice. For now branches of mathematics can be individuated by families of algorithmic systems: by (tacit) stipulation, one doesn't *change* mathematical practice; *new* mathematics is created by the introduction of new algorithmic systems (i) with rules different from all the others, and (ii) which aren't augmentations of systems already in practice. Should such an invention prove empirically applicable, and should it supplant some other (family of) system(s) previously applied to that domain, this doesn't cause a change in *mathematics*: the old family of systems is still mathematics, and is still something that can be profitably practiced (from the pure mathematical point of view). All that changes is the mathematics applied (and perhaps, the mathematics *funded*).

Notice that these reasons for the benign fixation of mathematical practice differ from those at work during mature mathematics. In particular, recall my thought experiment about the discovery of nonEuclidean geometry; its discovery, given when it happened, *spurred on* the detachment of mathematics (as a practice) from intended domains of application; but that was hardly something that it *started*. Mathematical development had already started to explode (in complex analysis, especially)—but although intended domains of application were still exerting a strong impact on the direction of mathematical research, the introduction of mathematical concepts was no longer solely a matter of abstracting and idealizing empirical notions, as the notion of the square root of -1 makes clear all on its own. I claim (but this is something only historians of mathematics can evaluate the truth of) that this,

(because the short-cuts used could presuppose inadmissible logical principles). It may be that this is correct: an adoption of a seriously deviant nonclassical logic for (some) mathematics requires formalization. I'm simply not sure: none of the factors that distinguish mathematical proof from formal derivation that I discuss in my 2004c seem to require any particular logic but my discussion there hardly exhausts such differences; and so I could easily be wrong about this.

39 What about classical number theory? Well, I'm not claiming that "mature" mathematics didn't have subject matters, the exploration of (some) of which wasn't expected to yield empirical application; but numbers aren't the best counterexample to my claim since they were clearly perceived to have (intended) empirical applications. The contemporary invention and exploration of whole domains of abstracta without (any) empirical application whatsoever is a different matter. Consider, e.g., most of the explorations of set-theoretic exotica or (all of) degree theory. (None of this is to say, of course, that empirical applications can't arise later.)

coupled with a more sophisticated view of how mathematical posits could prove empirically valuable (not just by a "resemblance" to what they're applied to), and both of these coupled with the emergence of a confident mathematical profession not directly concerned with the application of said mathematics, allowed the emergence of *mathematical liberalism*: the side-by-side noncompetitive existence of (logically incompatible) mathematical systems. And what a nice outcome *that* was!

ACKNOWLEDGMENTS

This paper owes its existence to an invitation by Jean Paul Van Bendegem and Bart Van Kerkhove to give a talk at a conference on Perspectives on Mathematical Practices (October 24-26, 2002). I also subsequently gave this talk at Columbia University on November 21, 2002. I'm grateful to both audiences for their suggestions which enabled improvements in the final version of the paper. I want to single out in this regard: David Albert, Haim Gaifman, and Philip Kitcher. Thanks also to Isaac Levi, Michael D. Resnik and Robert Thomas.

REFERENCES

Anderson, Stephen R. 1988. Morphological change. In *Linguistics: the Cambridge survey, Vol.I: Linguistic theory: foundations*. Cambridge: Cambridge University Press, 324-362.

Azzouni, Jody 1994. *Metaphysical myths, mathematical practice: the ontology and epistemology of the exact sciences*. Cambridge: Cambridge University Press.

------- 2000a. Applying mathematics: an attempt to design a philosophical problem, *The Monist* 83(2):209-227.

-------- 2000b. *Knowledge and reference in empirical science*. London: Routledge.

-------- 2004a. *Deflating existential commitment: a case for nominalism*. Oxford: Oxford University Press.

-------- 2004b. Proof and ontology in Euclidean mathematics. In *New Trends in the History and Philosophy of Mathematics*, eds. Tinne Hoff Kjeldsen, Stig Andur Pedersen, Lise Mariane Sonne-Hansen (pp. 117-33). Denmark: University Press of Southern Denmark.

-------- 2004c. The derivation-indicator view of mathematical practice, *Philosophia Mathematica* 12(3): 81-105.

Bloor, David 1983. *Wittgenstein: a social theory of knowledge*. New York: Columbia University Press.

Brown, James Robert 1999. *Philosophy of mathematics: an introduction to the world of proofs and pictures*. London: Routledge.

Descartes, René 1931a. Rules for the direction of the mind. In *The philosophical works of Descartes*, trans. by Elizabeth S. Haldane and G.R.T.Ross (pp. 1-77). Cambridge: Cambridge University Press.

-------- 1931b. Meditations on first philosophy. In *The philosophical works of Descartes*, trans. by Elizabeth S. Haldane and G.R.T.Ross (pp. 131-99). Cambridge: Cambridge University Press.

Dudley, Underwood 1987. *A Budget of trisections*. New York: Springer-Verlag.

Evans, Gareth 1973. The causal theory of names. In *Collected papers*, 1-24. Oxford: Oxford University Press (1985).

Hersh, Ruben 1997. *What is mathematics, really?* Oxford: Oxford University Press.

Jesseph, Douglas M. 1999. *Squaring the circle: the war between Hobbes and Wallis*. Chicago: University of Chicago Press.

Kline, Morris 1980. *Mathematics: the loss of certainty*. Oxford: Oxford University Press.

Kripke, Saul A. 1982. *Wittgenstein on rules and private language*. Cambridge, Massachusetts: Harvard University Press.

Orwell, George 1949. *1984*. New York: Harcourt, Brace and Company, Inc.

Putnam, Hilary 1975. The meaning of 'meaning'. In *Mind, language and reality: philosophical papers* (Vol. 2, pp. 139-52). Cambridge: Cambridge University Press.

Resnik, Michael D. 1997. *Mathematics as a science of patterns*. Oxford: Oxford University Press.

Westfall, Richard S. 1980. *Never at rest: a biography of Isaac Newton*. Cambridge: Cambridge University Press.

Wittgenstein, Ludwig 1953. *Philosophical investigations*. Trans. G.E.M. Anscombe. New York: The Macmillan Company.

Chapter 2

MATHEMATICS AS OBJECTIVE KNOWLEDGE AND AS HUMAN PRACTICE

Eduard Glas
Delft University of Technology

Abstract: Popper's world-3 doctrine is invoked to argue that characterizing mathematical developments as social processes is not incompatible with insisting on the objectivity and partial autonomy of mathematical knowledge. The argument is illustrated and supported by a historical case-study of the interplay between social and conceptual change in and after the French Revolution.

Key words: Mathematics, Popper, world-3, objectivity, autonomy, revolution, social and conceptual change

1. INTRODUCTION

Mathematics is the product of a communal human practice, but at the same time this product is partially independent of the practice that produced it. It possesses autonomous properties en relationships, which give rise to problems that are not merely human inventions. Although they are human creations, mathematical objects are not entirely transparent to their creators: we can make genuine discoveries of astonishing facts about our own creations.

These statements are easily recognizable as aspects of Popper's theory of objective knowledge. Popper saw mathematics – as well as science, art and other sociocultural institutions – as an evolutionary product of the intellectual efforts of humans. By objectivizing our creations and trying to solve the often unintended and unexpected problems that arise from those creations, we produce new mathematical objects, problems and critical arguments.

B. Van Kerkhove and J.P. van Bendegem (eds.), Perspectives on Mathematical Practices, 25–41.

Mathematical propositions are proved by logically compelling arguments, independently of empirical evidence, and therefore free of the interpretational ambiguities that make empirical scientific knowledge essentially uncertain. Real alternatives, in the sense of mutually incompatible theories generating conflicting claims about the truth value of particular statements, do not seem to exist in mathematics. Mathematicians may differ on many points, but when it comes to deciding about truth or falsity, it is proof alone that counts, and on that point debates are normally resolved by rational argumentation alone.

This situation has led many to assume that social and other external factors could not possibly play a constitutive part in the development of mathematical knowledge. Those, on the other hand, who insist on the essentially social nature of the mathematical enterprise – for instance social constructivists and adherents to the 'strong programme' – mostly begin by challenging some or all of the said characteristics.

I will argue that, in order to acknowledge the social dimension of mathematics, there is no need to question the objectivity and partial autonomy of mathematical knowledge in the Popperian sense. It is sufficient to shift our focus, away from the ways in which new truths are derived, towards the ways in which new problems are conceived and approached. There is indeed much more to mathematics than mere accumulation of true statements. Mathematicians are not interested just in truths (let alone truisms), but in truths that provide answers to questions that are worthwhile and promising in the contemporary scene of inquiry.

So, although there is no real problem of theory choice, as there is in science, there can be alternative choices of problems, and at this level there is a real possibility of competing schools with competing practices. To a certain extent, mathematical development is shaped by the conceptions of problems, aims and values that are shared by the members of a research community. Differentiation between communities of practitioners, with varying conceptions of what are relevant questions, appropriate ways of tackling them, and adequate criteria for appraising success, may therefore affect the development of a discipline. I will present a historical example that shows this evolutionary mechanism at work.

My presentation will consist of two parts. In the first part, I will discuss and defend Popper's theory of the evolution of objective mathematical knowledge, as an important alternative to the foundationist schools of formalism, intuitionism and platonism. For Popper, the theory of knowledge ultimately boils down to the theory of problem solving, and it is from this perspective that the sociocultural side of mathematics can fruitfully be approached, as I will try to show in the second part.

2. FALLIBILISM

Popper is not usually regarded as a philosopher of mathematics. As mathematical propositions fail to forbid any observable state of affairs, his demarcation criterion clearly divides mathematics from empirical science, and Popper was primarily concerned with empirical science. When speaking of a Popperian philosophy of mathematics, we mostly immediately think of Lakatos, who is usually considered to have applied and extended Popper's philosophy of *science* to mathematics. Like Lakatos, Popper saw considerable similarity between the methods of mathematics and of science – he held most of mathematics to be hypothetico-deductive (Popper 1984, p. 70) – and he thought highly of his former pupil's quasi-empiricist approach to the logic of mathematical development (Popper 1981, p. 136-7, 143, 165). His own views of the matter, however, are not to be identified with those of Lakatos, nor does their significance consist only in their having prepared the ground for the latter's methodological endeavours.

Popper never developed his views of mathematics systematically. However, scattered throughout his works, and often in function of other discussions, there are many passages which together amount to a truly Popperian philosophy of mathematics. This side of Popper's philosophy has remained rather underexposed, especially as compared with the excitement aroused by Lakatos's work, many of whose central ideas were developments of Popperian views, not only of science, but more specifically of mathematics as well (cf Glas 2001).

Already, in *Logik der Forschung*, Popper had argued that we should never save a threatened theoretical system by ad hoc adjustments, 'conventionalist stratagems', that reduce its testability (Popper 1972, p. 82-3) – a view which was to be exploited by Lakatos to such dramatic effect in the dialogues of *Proofs and Refutations*, under the heads of monster barring, exception barring, and monster adjustment. In *Conjectures and Refutations*, Popper had shown how the critical method can be applied to pure mathematics. Rather than questioning directly the status of mathematical truths, he tackled mathematical absolutism from a different angle. Mathematical truths may possess the greatest possible (though never absolute) certainty, but mathematics is not just accumulation of truths. Theories essentially are attempts at solving certain problems, and they are to be critically assessed, evaluated, and tested, by their ability to adequately solve the problems that they address, especially in comparison with possible rivals (Popper 1969, p. 197-9, 230).

This form of critical fallibilism obviously differed from Lakatos's quasi-empiricism, among other things by avoiding the latter's considerable problems with identifying the 'basic statements' that can act as potential falsifiers of mathematical theories. Even so, Lakatos's referring to what he called Popper's 'mistake of reserving a privileged infallible status for mathematics' (Lakatos 1976, p. 139 footnote) seems unjust. Claiming immunity to one kind of refutation – empirical – is not claiming immunity to all forms of criticism, much less infallibility. As a matter of fact, Popper did not consider anything, including logic itself, entirely certain and incorrigible (Popper 1984, p. 70-2).

3. OBJECTIVITY

Central to Popper's philosophy of mathematics was a group of ideas clustering around the doctrine of the relative autonomy of knowledge 'in the objective sense' – in contradistinction to the subjective sense of the beliefs of a knowing subject. Characteristic of science and mathematics is that they are formulated in a descriptive and argumentative language, and that the problems, theories, and errors contained in them stand in particular relations, which are independent of the beliefs that humans may have with respect to them. Once objectivized from their human creators, mathematical theories have an infinity of entailments, some entirely unintended and unexpected, that transcend the subjective consciousness of any human – and even of all humans, as is shown by the existence of unsolvable problems (Popper 1981, p. 161). In this sense, no human subject can ever completely 'know' the objective content of a mathematical theory, that is, including all its unforeseeable and unfathomable implications.

It is of course trivially true that knowledge in the said objective sense can subsist without anybody being aware of it, for instance in the case of totally forgotten theories that are later recaptured from some written source. It also has significant effects on human consciousness – even observation depends on judgements made against a background of objective knowledge – and through it on the physical world (for instance in the form of technologies). Human consciousness thus typically acts as a mediator between the abstract and the concrete, or the world of culture and the world of nature. To acknowledge that linguistically expressed knowledge can subsist without humans, that it possesses independent properties and relationships, and that it can produce mental and also – indirectly – physical effects, is tantamount to saying that it in a way exists. Of course, it does not exist in the way in which we say that physical or mental objects or processes exist: its existence is of a 'third' kind. As is well known, Popper coined the expression 'third world' (or 'world 3', as he later preferred) to refer to this abstract realm of objectivized products of human thought and language.

Popper's insisting upon the crucial distinction between the objective (third-world) and the subjective (second-world) dimension of knowledge enabled him to overcome the traditional dichotomies between those philosophies of mathematics that hold mathematical objects to be human constructions, intuitions, or inventions, and those that postulate their objective existence. His 'epistemology without a knowing subject' accounts for how mathematics can at once be autonomous *and* man-made, that is, how mathematical objects, relations and problems can be said in a way to exist independently of human consciousness *although* they are products of human (especially linguistic) practices. Mathematics is a human activity, and the product of this activity, mathematical knowledge, is a human creation. Once created, however, this product assumes a partially autonomous and timeless status (it 'alienates' itself from its creators, as Lakatos would have it), that is, it comes to possess its own objective, partly unintended and unexpected properties, irrespective of when, if ever, humans become aware of them.

Popper regarded mathematical objects – the system of natural numbers in particular – as products of human language and human thought: acquiring a language essentially means being able to grasp objective thought *contents*. The development of mathematics shows that with new linguistic means new kinds of facts and in particular new kinds of problems can be described. Unlike what apriorists like Kant and Descartes held, being human constructions does not make mathematical objects completely transparent, *clair et distinct*, to us. For instance, as soon as the natural numbers had been created or invented, the distinctions between odd and even, and between compound and prime numbers, and the associated problem of the Goldbach conjecture came to exist objectively: Is any even number greater than 2 the sum of two primes? Is this problem solvable or unsolvable? And if unsolvable, can its unsolvability be proved? (Popper 1984, p. 34). These problems in a sense have existed ever since humankind possessed a number system, although during many centuries nobody had been aware of them. Thus we can make genuine *discoveries* of independent problems and new hard facts about our own creations, and of objective (not merely intersubjective) truths about these matters.

Nothing mystical is involved here. On the contrary, Popper brought the platonist heaven of ideal mathematical entities down to earth, characterizing it as objectivized *human* knowledge. The theory of the third world at once accounts for the working mathematician's strong feeling that (s)he is dealing with something real, and explains how human consciousness can have access to abstract objects. As we have seen, these objects are not causally inert: for instance, by reading texts we become aware of some of their objective contents and the problems, arguments, etc., that are contained in them, so that the platonist riddle of how we can gain knowledge of objects existing outside space and time does not arise. Of course, speaking of causality here is using this notion in a somewhat peculiar, not in a mechanistic sense. That reading texts causes in us a certain awareness of what is contained in those texts is just a plain fact, for whose acceptance no intricate causal theory of language understanding is needed.

Cultural artefacts like mathematics possess their own partially autonomous properties and relationships, which are independent of our awareness of them: they have the character of hard facts that are to be *discovered*. In this respect they are very much like physical objects and relations, which are not unconditionally 'observable' either, but are only apprehended in a language which already incorporates many theories in the very structure of its usages. Like mathematical facts, empirical facts are thoroughly theory-impregnated and speculative, so that a strict separation between what traditionally has been called the analytic and the synthetic elements of scientific theories is illusory. The effectiveness of pure mathematics in natural science is miraculous only to a positivist, who cannot imagine how formulas arrived at entirely independently of empirical data can be adequate for the formulation of theories supposedly inferred from empirical data. But once it is recognized that the basic concepts and operations of arithmetic and geometry have been designed originally for the practical purpose of counting and measuring, it is almost trivial that all mathematics based on them remains applicable exactly to the extent that natural phenomena resemble operations in geometry and arithmetic sufficiently to be conceptualized in (man-made) terms of countable and measurable things, and thus to be represented in mathematical language. In mathematics and physics alike, theories are often put forward as mere speculations, mere possibilities, the difference being that scientific theories are to be tested directly against empirical material, and mathematical theories only indirectly, if and in so far as they are applied in physics or otherwise (Popper 1969, p. 210, 331).

4. INTERACTION

It is especially the (dialectic) idea of *interaction* and partial *overlap* between the three worlds that makes Popper's theory to transcend the foundationist programmes. Clearly, objective knowledge (at world-3 level) – the objective contents of theories – can exist only if those theories have been materially realized in texts (at world-1 level), which cannot be written nor be read without involving human consciousness (at world-2 level). Put somewhat bluntly, platonists acknowledge only a third world as the realm to which all mathematical truths pertain, strictly separated from the physical world; intuitionists locate mathematics in a second world of mental constructions and operations, whereas formalists reduce mathematics to rule-governed manipulation with 'signs signifying nothing', that is, mere material (first-world) 'marks'. In all these cases, reality is split up into at most two independent realms (physical and ideal or physical and mental), as if these were the only possible alternatives. Popper's tripartite world view surpasses physicalist or mentalist reductionism as well as physical/mental dualism, emphasizing that there are *three* partially autonomous realms, intimately coupled through feed-back. The theory of the interaction between all three worlds shows how these seemingly incompatible mathematical ontologies can be reconciled and their mutual oppositions superseded (Popper 1984, p. 36-37; cf Niiniluoto 1992).

The notion of a partially autonomous realm of objective knowledge has been criticized, most elaborately by O'Hear in his Popper monograph (O'Hear 1980). O'Hear does not deny that objectivized mathematical theories have partly unforeseeable and inevitable implications, but he does not consider this sufficient reason for posing what he calls 'an autonomous non-human realm of pure ideas'. Popper, of course, always spoke of a *partially* autonomous realm, not of 'pure ideas', but especially of fallible theories, problems, tentative solutions and critical arguments. O'Hear, however, argues that Popper's theory is misleading because it implies that we are not in control of world 3 but are, on the contrary, completely controlled *by* it (*ibid.*, p. 183, 207). For relationships in world 3 are of a *logical* character and this seems to imply that they are completely beyond our control. On O'Hear's construal, Popper allowed only a human-constructive input at the very beginning of the history of mathematics – the phase of primitive concepts connected with counting and measuring – after which logic took over and developments were no longer under human control. World 3 would be entirely autonomous rather than only partially autonomous, and mathematicians would be passive analyzers rather than active synthesizers of mathematical knowledge – almost the opposite of Popper's earlier emphasis on the active role of the subject in observation and theory formation. I think that these conclusions rest on a misunderstanding of the logical character of relationships in world 3.

To stress the objective and partly autonomous dimension of knowledge is not to lose sight of the fact that it is created, discussed, evaluated, tested and modified by human beings. Popper regarded world 3 above all as a product of intelligent human practice, and especially of the human ability to express and criticize arguments in language. The objectivity of mathematics rests, as does that of all science, upon the criticizability of its arguments, so on language as the indispensable medium of critical discussion (Popper 1981, p. 136-137). Indeed, it is from language that we get the idea of 'logical consequence' in the first place, on which the third world so strongly depends. But mathematics is not *just* language, and neither is it *just* logic: there are such things as extra-logical mathematical objects. And although critical discussion depends on the use of discursive language, mathematics is not bound to one particular *system* of logic. O'Hear (1980, p. 191-198) rightly argued that there is room for choices to fit our pre-systematic intuitions and even physical realities (he for instance discusses deviating logics to fit quantum mechanics). But the possibility of alternative logics does not invalidate the idea of logical consequence as such, it does not make one or the other of alternative logical systems illogical. The choice of a *specific* logical system for mathematics or science has itself to be decided by 'logical' argumentation (in the *general* sense of the term).

Although the third world arises together with argumentative language, it does not consist exclusively of linguistic forms but contains also non-linguistic objects. As is well known, the concept of number, for instance, can be axiomatically described in a variety of ways, which all define it only up to isomorphism. That we have different logical explications of number does not mean that we are talking about different objects (nor that the numbers with which our ancestors worked were entirely different from ours). We must distinguish between numbers as third-world objects and the fallible and changing theories that we form about these objects. That the third-world objects themselves are relatively autonomous means that our intuitive grasp of them is always only partial, and that our theories about them are essentially incomplete, unable to capture fully their infinite richness.

5. POPPERIAN DIALECTIC

The idea that the third world of objective mathematical knowledge is partly autonomous does not at all imply that the role of mathematicians is reduced to passive observation of a pre-given realm of mathematical objects and structures – no more than that the autonomy of the first world would reduce the role of physicists to passive observation of physical states of affairs. On the contrary, the growth of mathematical knowledge is almost entirely due to the constant feed-back or 'dialectic' between human creative action upon the third world and the action of the third world upon human thought. Popper characterized world 3 as the (evolutionary) product of the rational efforts of humans who, by trying to eliminate contradictions in the extant body of knowledge, produce new theories, arguments, and problems, essentially along the lines of what he called 'the critical interpretation of the (non-Hegelian) dialectic schema: $P_1 \rightarrow TT \rightarrow EE \rightarrow P_2$' (Popper 1981,

p. 164). P_1 is the initial problem situation, that is, a problem picked out against a third-world background. TT is the first tentative theoretical solution, which is followed by error elimination (EE), its severe critical examination and evaluation in comparison with any rival solutions. P_2 is the new problem situation arising from the critical discussion, in which the 'experiences' (that is, the failures) of the foregoing attempts are used to pinpoint both their weak and their strong points, so that we may learn how to improve our guesses.

Every rational theory, whether mathematical or scientific or metaphysical, is rational on Popper's view exactly 'in so far as it tries to solve certain problems. A theory is comprehensible and reasonable only in its relation to a given problem situation, and it can be discussed only by discussing this relation' (Popper 1969, p. 199). In mathematics as in science, it is always problems and tentative problem solutions that are at stake: 'only if it is an answer to a problem – a difficult, a fertile problem, a problem of some depth – does a truth, or a conjecture about the truth, become relevant to science. This is so in pure mathematics, and it is so in the natural sciences' (*ibid.*, p. 230). Popper clearly did not view mathematics as a formal language game, but as a rational problem solving activity based, like all rational pursuits, on speculation and criticism.

Although they have no falsifiers in the logical sense – they do not forbid any singular spatiotemporal statement – mathematical theories (as well as logical, philosophical, metaphysical and other non-empirical theories) can nevertheless be critically assessed for their ability to solve the problems in response to which they were designed, and accordingly improved along the lines of the *situational* logic or dialectic indicated above. In particular, mathematical and other 'irrefutable' theories often provide a basis or framework for the development of scientific theories that *can* be refuted (Popper 1969, chapter 8) – a view which later was to inspire Lakatos's notion of research programmes with an 'irrefutable' hard core (Lakatos 1978, p. 95).

Most characteristic of Popper's approach to mathematics was his focussing entirely on the dynamics of conceptual change through the dialectic process outlined, replacing the preoccupation of the traditional approach with definitions and explications of meanings. Interesting formalizations are not attempts at clarifying meanings but at solving problems – especially eliminating contradictions – and this has often been achieved by *abandoning* the attempt to clarify, or make exact, or explicate the intended or intuitive meaning of the concepts in question – as illustrated in particular by the development and rigorization of the calculus (Popper 1983, p. 266). From his objectivist point of view, epistemology becomes the theory of problem solving, that is, of the construction, critical discussion, evaluation, and critical testing, of competing conjectural theories. In this, everything is welcome as a source of inspiration, including intuition, convention and tradition, especially if it suggests new problems. Most creative ideas are based on intuition, and those that are not are the result of criticism of intuitive ideas (Popper 1984, p. 69). There is no sharp distinction between intuitive and discursive thought. With the development of discursive language, our intuitive grasp has become utterly different from what it was before. This has become particularly apparent from the twentieth-century foundation crisis and ensuing discoveries about incompleteness and undecidability. Even our logical intuitions turned out to be liable to correction by discursive mathematical reasoning (*ibid.* p. 70).

6. SOCIALLY CONDITIONED CHANGE

So, mathematics is primarily conceived as a problem solving practice, and – as Popper explicitly stated – anything is welcome as a source of inspiration, especially if it suggests new problems. I will now briefly discuss a case of *socially* conditioned mathematical change, that is, a case in which social processes were the main sources of conceptual innovation, and argue that it is perfectly well possible to acknowledge the social nature of the mathematical enterprise without denying its objectivity and partial autonomy (for a historically detailed account, see Glas 2002).

Among the important driving forces of mathematical development are concrete, often scientific or technological problems. The calculus, for instance, was developed primarily as an indispensable tool for the science of mechanics. Mathematicians were well aware of its lack of rigor and other fundamental shortcomings, but its impressive problem solving power was reason enough not to abandon it. Instead, eighteenth-century mathematicians tried to perfect the calculus by detaching it from its geometric roots and reformulating it as a linguistic system based on deductions from proposition to proposition, without any appeals to figure-based reasoning. This was achieved by interpreting variables as non-designated quantitites and by introducing the notion of function, which replaced the study of curves (cf Ferraro 2001). In the last quarter of the century, Condillac's view that language was constitutive of thought, and the *langue des calculs* its highest manifestation, was shared by many intellectuals, among them the most prominent French mathematicians of the time, Lagrange and Laplace (cf Glas 1986, p. 251- 256).

A new chapter in the history of mathematics began when the new, machine-driven industrial technologies gave rise to an entirely new type of problems, and with it to a new mathematical approach, which entered into competition with the established analytical doctrine. It was standard procedure in analytical mechanics to deduce from the principles of mechanics the particular rules of equilibrium and motion in such devices as the lever, the crank, and the pulley. These 'machines', however, were idealized to the point of ignoring all material aspects of real technical devices. The approach was therefore of very limited use to the actual practice of mechanical engineering. Engineers still worked mainly with empirical rules of thumb based on trial and error. This problem – supplying an exact scientific foundation to the practice of engineering – lay at the root of the new course that mathematics embarked on.

The founders of this new approach to mathematics were Carnot and Monge, who both developed new forms of geometry concurrently with their involvement with engineering problems. Before the revolution they were not highly regarded as mathematicians, and without the military needs of the revolution they could scarcely have stood up to the competition of the leading analytical and anti-geometric style of thinking. It was their personal engagement in revolutionary politics that eventually put Carnot and Monge in a position to carry through a radical educational reform – embodied in the Ecole Polytechnique – which was essential to the formation of a new community of mathematically versed engineers (*ingénieurs savants*), who shared their particular views of the problems, aims and methods of mathematics.

Carnot initiated the science of machines as a domain in itself, and developed a new geometry concurrently with it. Finding the operational principles of machines required a new conception of geometry, less static and figure-bound than the classical version, not concerned with the fixed properties of immutable forms but with the much more general problem of possible movements in spatial configurations. The new geometry was focussed on discovering the general principles of transformation of spatial systems rather than on deducing the properties of particular figures from a set of pre-established principles (see Gillispie 1971).

Carnot's work (Carnot 1783) did not read like the eighteenth-century analytical mechanics that culminated in Lagrange's *Mécanique analytique* (Lagrange 1811 [1788]). It in fact appears barely to have been read at all and its author was not known as a scientist before the revolution. His work presupposed the competence of persons versed in abstract scientific thought, yet was written in a geometric idiom that was not suited to arouse the interest of mathematicians, who were primarily concerned with the further perfection of analysis by purging it of all remnants of figure-based reasoning. It apparently was addressed at scientifically versed engineers like himself, an intended audience which at the time existed, if at all, only *in statu nascendi.*

As an officer of the army's engineering corps (*génie*), Carnot had been educated at the military *Ecole du Génie* (School of Engineering) at Mézières, where his fellow revolutionary Monge had been his teacher. The latter's *Géométrie descriptive* (Monge 1811 [1799]) originated in the same practical engineering context and it likewise gathered geometric subject matter under a point of view at once more general than classical geometry and operational in tackling engineering problems.

In Monge's new approach to geometry, the objects of research were not defined by the particular *forms* of geometric figures, but by the *methods* used for generating and interrelating spatial configurations. Apart from pure, projective geometry, descriptive geometry furnished the basis of major developments in analytical geometry, differential geometry and pure analysis (cf Glas 1986, pp. 256-261). The constant association of analytical expressions with situations studied in geometry – on which Monge placed so high a pedagogic value – was of the greatest consequence to the image of mathematics as a whole and set the stage for what we now call 'modern' geometry. It was the particular combination of synthetic and analytic qualities, bringing analysis and geometry to bear on each other in entirely new ways, that made Monge's approach truly novel and accounts for the fecundity of its leading ideas in the exploration of various new territories.

Monge himself linked the aim and object of his geometry directly to its indispensability as a language-tool for modern industrial practice, based on division of labour and therefore having to rely on cooperation and communication between all the heads and hands involved in technical-industrial projects. Besides providing forceful means of tackling problems of design, construction and deployment of machines with the mathematical precision required by the new industrial technologies, descriptive geometry would serve as the indispensable common language of communication between all participants in the productive order of society, who otherwise would remain divided by boundaries of class, profession, and function.

Monge and Carnot both developed their mathematics concurrently with their involvement with engineering problems. Paradoxically, despite the practical and applied nature of the problems that they envisaged, the mathematics that they developed on this basis stands out by its generality and purity. It was not of a lesser standard than the authoritative analytical approach, but was oriented towards different cognitive aims: integrating formal and functional features of spatial systems rather than deducing the consequences of pre-established principles. Like Carnot's science of geometric motion, Monge's descriptive geometry was the intellectual response to the new problems, connected with the rise of machine-driven industrial technologies, that faced their professional community.

7. ALTERNATIVE PRACTICES

Whatever may be thought of the intellectual value of Monge's and Carnot's contributions to mathematics, this is certainly not a case of 'superior minds' bending the course of an entire discipline by force of reason alone. Their intellectual achievements were inextricably tied up with their professional, social, and political engagement, which the vicissitudes of the revolution allowed them to put into effect by the creation of the Ecole Polytechnique. It was through this educational reform that their work became exemplary for a whole new generation of mathematicians, who under their inspiration opened up fresh and fertile fields of inquiry (whereas the analytical 'language' view made mathematics to appear very nearly completed).

The great changes in mathematics that the birth of the new community engendered – Boyer, for instance, speaks of a geometric and an analytical 'revolution' (Boyer 1968, p. 510) – are only understandable in virtue of the institutional, social and political development of the profession of engineering. This is not to say that the causal arrow points only in one direction, from the social to the intellectual. The intellectual and the social developments were mutually constitutive, but in this particular (perhaps exceptional) case the changed socio-political and institutional role and organization of the professional community of engineers should at least be given explanatory priority as *conditio sine qua non*. For without these developments, the conceptual innovations that Monge and Carnot carried through would have missed their target; they would not have found an appreciative audience and could scarcely have had any impact on the course of mathematical development, as evidenced by their almost total failure to make any impression on the rest of the scientific world prior to the revolution.

The case should certainly not be reduced to a simple conflict of interests between separate specialties, geometry and analysis. Although the followers of the analytical main stream were in a sense 'against' geometry (to the point of making the discipline nearly extinct), Carnot and Monge were not at all 'against' analysis, quite the contrary. Like their analytical colleagues, they considered it ideal for the representation and calculation of variations, and therefore indispensable for any engineer. Carnot made abundant use of analysis, and Monge even contributed considerably to its progress. We in fact owe the modern 'analytical geometry' in large measure to Monge's purely analytical characterization of lines, surfaces and solids in space.

Monge and Carnot placed themselves outside the ruling analytical tradition, not because they were geometers rather than analysts, but because they found themselves confronted with altogether different sorts of problems. They were not so much concerned with the further 'linguistic' perfection of analysis as with the concrete problems that faced their own professional (engineering) community, problems that could not be handled adequately by the contemporary analytical mechanics. The geometry that they developed in response to these problems differed fundamentally from the classical version in that its objects were defined in terms of operations and transformations, not in terms of particular types of figures. The main reason for eighteenth-century mathematicians to abandon geometry had been its relying on figure-based reasoning instead of logical deduction. Rather than joining in with this general rejection, Carnot and Monge developed a new conception of geometry altogether, detaching it from consideration of particular figures and redefining it as the study of spatially extended structures and their transformations. Although of course figures figured prominently in Carnot's and Monge's works, their mode of reasoning was not figure-bound. Figures were just heuristic means of investigation, useful to direct and support the geometric reasoning, which in itself proceeded at a level of abstraction and generalization that went far beyond what could be represented figuratively. Their conceptual innovation made the classical distinctions between analytic and synthetic methods obsolete; indeed, it was precisely the intimate unity of synthetic and analytic reasoning that made their approach truly novel and fruitful.

Under the *Empire*, the mathematicians of the analytic tradition regained much of the territory they had lost to the geometric innovators in the revolutionary days. Laplace in particular, who had not been 'seen' by the revolutionaries, was highly regarded by the emperor Napoleon, who clearly was more sensitive to the value of the 'old' tradition in point of respectability and prestige (cf Bradley 1975). The leading role of Laplace in the 'imperial' reformation of the Polytechnic is reflected in the changing relative positions of geometry and analysis in its programme. The time tables of the school show how geometry dropped from 50 hours in 1795 to 27.5 in 1812, whereas analysis and analytical mechanics in this period rose from 8 to 46, also to the cost of chemistry (Dhombres 1989, p. 572). But the turning of the tide, and even the eventual expelling of Carnot and Monge from the scientific institutions of France under the Restoration, could not make undone what had happened to mathematics. It had integrated the spirit of the revolutionary method and had become itself an important element of intellectual and social change.

Until deep into the nineteenth century, a remarkably sharp division subsisted between two groups of mathematicians, the one taking its inspiration from Carnot and Monge and primarily motivated by constructive problems in a technological context, the other more in line with Lagrange and Laplace and chiefly concerned with analytical problems in a general scientific setting (cf Grattan-Guinness 1993, pp. 408-411, who lists seventeen mathematicians in each group).

8. CONCLUDING DISCUSSION

The case underscores the explanatory priority of communal practices in accounting for a type of conceptual change in mathematics that is fundamental to its advance. It was their belonging to a different professional group that accounts for the different viewpoints, problems, aims, values, methods, and approaches of Carnot and Monge, as compared with the received analytical tradition. There never was disagreement about the particular contents of mathematical theories, the correctness of theorems and proofs, and the like. But the two schools differed fundamentally on such issues as what were the questions most worth asking, the methods most appropriate for handling them adequately, and the right criteria for appraising progress. There was no crisis in the extant doctrine, no accumulation of insuperable problems that demanded an entirely new approach. At most it can be said that the reigning tradition had largely exhausted itself: it was regarded as completed, and interesting new discoveries were no longer considered possible. The revolution was not sparked off by deep epistemological worries that led to a replacement of theories; instead, it has to be characterized as a replacement of research communities, the emergence of a new community of mathematically versed engineers, setting themselves radically different sorts of problems, and demanding different methods to solve them.

As a social process, this particular revolution in mathematics shared some of the features of a Kuhnian revolution, but in other respects it was quite different. The practices of the old and the new 'school' were incommensurable by being *at cross-purposes*, rather than in disagreement about the truth or falsity of each other's results. There was no problem of theory choice – the central problem in Kuhn's account of scientific revolutions – but a fundamental shift of evaluative standards, which accounts for the impossibility of resolving the differences by logical argumentation alone. The change was not induced by serious epistemological problems, but by a radical change in the social conditions under which mathematicians worked, the most significant sign of which was the moving of the (military) engineers – with their characteristic view of the objects, problems, aims and values of mathematics – to the centre of state power.

The case shows that in order to characterize mathematical change as a social process, there is no need to question the objectivity and rationality of mathematics in the sense that Popper gave to these notions. A shift of focus, away from the ways in which new truths are derived, and towards the ways in which new problems are conceived and handled, is sufficient. Indeed, unlike the Hegelian dialectic, the Popperian dialectic does not start off with theses but with problems. Problems are the initial and the central motives in the development of mathematics, which is understood primarily as a problem solving practice. It is the dynamics of problem situations, rather than the statics of definitions and theorems, that is characteristic of the growth of mathematical knowledge.

Mathematics is a social practice, shaped by the ways in which its practitioners view the problems that face their community. It is through language that mathematicians can lay out their thoughts objectively, in symbolic form, and then develop, discuss, test and improve them. Humankind has used descriptive and argumentative language to create a body of objective knowledge, stored in libraries and handed down from generation to generation, which enables us to profit from the trials and errors of our ancestors. Characteristic of science and mathematics is that they are formulated in objective language, and that the problems, theories, and errors contained in them stand in logical relations that – *pace* Kuhn – are independent of individual or collective beliefs and other mental states that humans may have with respect to the contents involved.

REFERENCES

Boyer, C. B., 1968, *A History of Mathematics* (New York: Wiley)

Bradley, M., 1975, 'Scientific Education versus Military Training: The Influence of Napoleon Bonaparte on the Ecole Polytechnique', *Annals of Science* **32**, 415-449.

Carnot, L. N. M., 1783, *Essai sur les machines en général* (Dijon: no publisher)

Dhombres, J. & Dhombres, N., 1989, *Naissance d'un nouveau pouvoir: sciences et savants en France, 1793-1824* (Paris: Payot)

Ferraro, G., 2001, 'Analytical Symbols and Geometrical Figures in Eighteenth-Century Calculus', *Studies in History and Philosophy of Science* **32**, 535-555

Gillispie, C. C., 1971, *Lazare Carnot Savant* (Princeton: Princeton University Press)

Glas, E., 1986, 'On the Dynamics of Mathematical Change in the Case of Monge and the French Revolution', *Studies in History and Philosophy of Science* **17**, 249-268

Glas, E., 2001, 'The Popperian Programme and Mathematics', *Studies in History and Philosophy of Science* **32**, 119-137, 355-376

Glas,E., 2002, 'Socially Conditioned Mathematical Change: The Case of the French Revolution', *Studies in History and Philosophy of Science* **33**, 709-728

Grattan-Guinness, I., 1993, 'The ingénieur-savant, 1800-1830: A Neglected Figure in the History of French Mathematics and Science', *Science in Context* **6**, 405-433

Lagrange, J. L., 1811, *Mécanique analytique*, nouvelle édition, orig. 1788 (Paris: Courcier)

Lakatos, I., 1976, *Proofs and Refutations: The Logic of Mathematical Discovery*, ed. J. Worrall and G. Currie (Cambridge: Cambridge University Press)

Lakatos, I., 1978, *The Methodology of Scientific Research Programmes* (Philosophical Papers Vol.1), ed. J. Worrall and G. Currie (Cambridge: Cambridge University Press)

Monge, G., 1811, *Géométrie descriptive*, nouvelle édition, orig. 1799 (Paris: Klostermann)
Niiniluoto, I., 1992, 'Reality, Truth, and Confirmation in Mathematics – Reflections on the Quasi-Empiricist Programme', pp. 60-77 in Echeverria, J., Ibarra, A. and Mormann, T. (eds.), *The Space of Mathematics* (Berlin, New York: De Gruyter)
O'Hear, A., 1980, *Karl Popper* (London: Routledge & Kegan Paul)
Popper, K. R., 1969, *Conjectures and Refutations: The Growth of Scientific Knowledge*, third edition (London: Routledge)
Popper, K.R., 1972, *The Logic of Scientific Discovery*, sixth edition (London: Hutchinson)
Popper, K. R., 1981, *Objective Knowledge: An Evolutionary Approach*, revised edition (Oxford: Clarendon)
Popper, K. R., 1983, *Realism and the Aim of Science*, ed. W. W. Bartley (Totowa: Rowen and Littlefield)
Popper, K. R., 1984, *Auf der Suche nach einer besseren Welt* (München: Piper)

Chapter 3

THE COMPARISON OF MATHEMATICS WITH NARRATIVE

R. S. D. Thomas
University of Manitoba

Abstract: Mathematical writing, chiefly of proofs, is compared with the telling of stories. Contrasts are also noted. The positive analogy is used to support the view of mathematics as being about relations rather than objects obviating a need for ontological commitment to mathematical objects. The negative analogy is used to deny some philosophers' identification of mathematics with fiction.

Key words: Narrative, fiction, history, relations, pretence

1. INTRODUCTION

Once upon a time, a foreign academic and his wife were spending a couple of days as tourists in Cambridge adjacent to his attending a conference when they saw a notice of a public lecture in a venue that aroused their curiosity—one of those places that you have to pay to see if you are not there for a purpose. It is always pleasant as tourists to have somewhere attractive to sit down, and so they went to the lecture and found it as interesting as the venue. The lecturer handed out a sheet of paper with a summary of the text that the lecture was meant to discuss, a bit of background, an example, and some bibliography. When the academic got back home, he followed up some of the references and hatched a research project that could not have been more directly inspired by the public lecture if it had been about his own subject, which it was not.

That little story is not, so far as I know, a true story; I have made it up to be fictional but to be applicable to my encounter with narrative in Oxford in

B. Van Kerkhove and J.P. van Bendegem (eds.), Perspectives on Mathematical Practices, 43–59.

1997 by a process, which we all understand and which Mark Turner, for whom it is important, calls projection [1996]. I want also to give you the sequel.

Having found that narrative was something that was studied, something I should have guessed if I had ever thought about it, I thought that I would use that fact to tackle the puzzling comparison I had seen in several places between mathematics and fiction, most fiction nowadays being in narrative form. I thought that to do so would take me a few weeks, and then I could get on to some other project. A year later I was not finished, but my sabbatical had come to an end. What delayed me so much?[1] Partly the large number of persons that have made usually superficial comparisons of mathematics with fiction, not all of them saying the same thing. Partly that, wherever I turned I found, whether mentioned in the literature or not, similarities between proofs and narratives, theorems and stories. There turned out to be so much material here that, after I had organized what I thought I wanted to say about it, I encountered John Allen Paulos's book, *Once upon a number* [1998], in which he does something rather similar written at about the same time, comparing stories and statistics with hardly any overlap with what I had found. What am I talking about? If one has not thought about this, it seems downright odd. So let me spell out what I mean.[2]

2. STORIES

I begin by indicating what about stories makes the comparison work. Most important is the initial statement of some relations among characters and perhaps props that are the focus of the story. The beginning of a story is signalled somehow, for example by 'once upon a time' or the proscenium arch. Like its beginning, a story ends with a situation involving the characters, whether all dead or 'they lived happily ever after'. A story almost invariably engages a listener's attention and imagination in a way that other speech acts do not, wills perhaps excepted. A story about a mother and a child is more interesting than an essay on motherhood or childhood without examples; the relations in a story are exemplified rather than reified and named. The world of the story is what the listener makes of the tale in co-operation with the teller. While it may be imaginary, it often bulges with

[1] I gratefully acknowledge the hospitality of Wolfson College Oxford and the Philosophy Centre, University of Oxford, during what I have characterized as delay above and again while I have prepared this document. I am also grateful to my department head during these years and those between for his lack of complaint at my being side-tracked into doing philosophy from my more normal activities of editing and mathematics.

[2] I have done this for mathematicians in [2002], to which the next section is very similar.

authenticating detail. The usually invented relation instances in a story, unlike the invented characters, are almost always ordinary relations in the real world, even in science fiction. In spite of contributing heavily to the world of the story, different hearers will retell the story recognizably. Stories have greater integrity than non-narrative rumours. While the listener imagines the story, the listener does not make it up. There is some evidence that the opposite is true: one is, to some extent, the stock of one's imagination, the product of the stories that one knows, including—of course—one's own. The stories we know tell us how things can be in the world; through them, we see the world, though not necessarily as it is. The important characteristics of characters are mainly relational and are given to them by the storyteller. A most economical way of conveying a lot about characters is by using actors to stage a play. I am talking about plot-driven narrative rather than stories that concentrate on development of character and seeing inside characters' heads as began to be fashionable in novels two centuries ago. I'm thinking of simple stories like fairy tales, murder mysteries, and much science fiction.

A story is not just descriptions in any order, but rather the tale proceeds in accordance with physical causality moved along by characters' reasons and intentions. A story is structured, not just a list of who was where when, and turning the one into the other is hard work. A list of facts about the basket, the grandmother, the little girl, and the wolf do not convey 'Little Red Riding Hood' even to people who already know the story according to Adam Morton on the basis of experiment [1996]. A listener needs logic as well as imagination; each new remark, character, and event needs to be fitted into the structure the listener has already drawn in mental space.[3] This multi-faculty engagement is much of the pleasure of hearing a story. Telling a story, however complex, requires presentation of plots and subplots in a linear way with devices like flashbacks to fill in out-of-order details. Part of the art of storytelling is in emphasizing the good bits, building tension, and resolving conflict in a satisfying way that I fortunately need not describe.

There is more than one way to look at the world; fortunately we know more than one story about the same relations, sometimes even about the same historical characters. Story material can be applied to the real world, for example, in my Cambridge tale, or when we say history is repeating itself, or in saying someone is a Scrooge or Uriah Heep; there are famous fictional as well as historical behaviours.

The truths of fiction are not those of the newspaper but revelations of a higher order (the moral of the story). When we discuss what happened in a

[3] This matter, the logic of fiction, has been the subject of much discussion. See Eco [1979] and Woods [2002] for very different approaches---and my last paragraph.

story, we return to the text, which is the touchstone of the story's objectivity. There can be serious discussion about the subject matter of a fictional story. Fans talk, indeed argue, about Sherlock Holmes without telling a Holmes story, but their discussion depends on those stories.[4] We may ask, in such discourse, how fictional characters are created, or we may ask 'factual' questions like 'how long is Hamlet's nose?' Such 'fictional/factual' questions may not be answerable if neither the text nor deduction from it tells us. Finally, stories are important and ubiquitous; some fictional and many historical characters are better known than almost all of our contemporaries.

Note that my list of characteristics of stories was tedious (if strange) in part because I could give it neither narrative nor logical structure. It forms a contrast with the story with which I began. In indicating how mathematical discourse is similar to narrative, I shall base the order of my discussion on that of my discussion of stories, arbitrary though that was.

A mathematical proof begins with our postulating some things to talk about, the relations that they have among themselves, and the symbols that we shall use. Everyone recognizes this move as a mathematical beginning once they have learned to let x be the number of coins in Johnny's pocket. We frequently begin by calling upon a standard set of objects and relations provided by an axiomatic system. This approach is analogous to telling a new story about characters already known from their appearance in the local mythology: Brunhilde is the daughter of Wotan, *etc*. What follows on from such relations in mathematics—actualizing potential, as it is called in stories—is primarily deductive. Time passes, but the time is the reader's; and the presentation is more or less linear, as in a story. A proof ends with another set of relations among the characters, it being a matter of choice where to end the deductive chain, since most mathematical situations have further logical consequences. This choice is not made for logical reasons but defines the conclusion of the theorem.

While fiction and history focus on significant relations among characters, accidents, chance encounters, and other external events are often inserted into narrative either, in fiction, as authenticating detail or, in history, because they happened. Mathematics, in contrast, is supposed to depend only upon the stated relations among the mathematical objects; the importation of other relations would be the opposite of authentication, however intuition may guide deciding what to prove and how. A feature of fiction that is not obvious to readers but is well-known to writers is a fictional analogue of mathematical deduction. In spite of their having invented everything in the

4 For this reason it has been suggested (van Inwagen [1977]) that, corresponding to each (non-existent) fictional character there is created an (existent) abstract object.

story except human and physical nature, it must work itself out. In both cases discovery is experienced in spite of its basis in invention. Intuition is important for such discoveries, whether in mathematics or fiction or perhaps history.

All physical causes, personal intentions, and logical consequences in stories are mapped to implication in mathematics. All are answers to the implicit question, 'how does it turn out?'. Completely unjustified mathematical facts are too boring to keep awake for and almost impossible to remember. The parts played by imagination and deduction in stories and mathematics are interestingly different, indeed almost opposite. In mathematics one imagines in order to understand implications, whereas in stories one deduces locally to know how to imagine the story's course. Deduction does not work on a large scale in reading for the same reason as in meteorology; in stories as in the real world there are too many borderline cases and hidden variables for deduction over any distance to be dependable. The lack of closure is one of the matters discussed in the logic of fiction.

As in stories, in mathematics one sees possibilities and—unlike stories—impossibilities. As well as learning to see in mathematical terms, one also learns that some relations are not possible in the presence of others. Fiction is more purely permissive on account of proceeding by describing.

An engaging feature of mathematical discourse is that the relations discussed are actually relating, not abstracted and reified. Geometry is silent on collinearity as an abstraction; our study of collinearity and non-collinearity concerns points that are collinear or are vertices of polygons. With three non-collinear points we have a triangle and can calculate its area, reasoning about the vertices' non-collinearity. Shakespeare has occasional remarks on abstractions like jealousy and treachery, but is famous not for essays on jealousy and treachery, but for creating Othello and Macbeth to embody them. Even the most abstract mathematics has entities *in* the relations being discussed. H. B. Griffiths has pointed out to me that specializing an algebraic structure, distinguishing one specific example of a class of isomorphic structures, 'is like casting a play, and the flavour of the special mathematics corresponds to that of a particular production: all such productions have the same abstract structure.'

We draw general conclusions from consideration of cases that we take care are not *special cases*, what we call *general cases*, a linguistic oddity. We can eliminate accidental characteristics by using the language of set theory, in which the objects are completely characterless and the relations that we imagine are all specified in terms of a few simple relations like set membership and set inclusion. Oppositely, narrative discusses what are frankly and *ultimately* special cases. This is not, though in one way opposite, altogether different; mathematicians are also interested in special cases for the patterns they reveal, and the best of narrative's special cases have a similar purpose that is left implicit, which is why Scrooge and Heep become paradigms.

Mathematicians postulate simple relations and only those needed; there is no concern with fully rounded characters. But they are discussed as though they were real with ordinary semantics just as storytellers talk the same way whether recounting history or fiction.

The imaginative effort required to learn a mathematical proof is typically considerable, the reader makes a contribution. If the learner then repeats the proof, most of what was imagined will be ignored, and the proof given will be substantially the same. Knowing just what and how much to imagine was part of the learning. We somehow grasp a proof as a whole as we somehow grasp a story as a whole. This has led Brian Rotman to write [1993] of what he calls the *story* of the proof. There are aspects of presentation that help with the grasping. Stories and proofs as encountered are linear; we have lemmas that allow us to prove things out of order like flashbacks. We define certain results to be of greater importance and specify them as theorems or lemmas for that reason; in a novel they might be chapters. There is an analogue in proof of dramatic tension, as there is of its elegant release.

The importance of the text for the objectivity of mathematics, while less than in fiction, has led to the philosophical position—extreme formalism—that the text of proofs is all there is. Just as we can entertain more than one story about some mythological characters, we can welcome, for example, different formulations of the positive integers within set theory and different proofs of the same result. I have already hinted that mathematics can be applied in the same way as stories. Triangles are among the few things that occur both physically in the real world and metaphorically in stories. For the most part, however, we are not interested in the application to single things of mathematical names but rather in the application of whole theories like Euclidean space to whole scientific theories like Newtonian mechanics. Then what corresponds to a triangle in the mathematics is typically not a physical triangle but just three non-collinear things, as in the three-body problem. The objects of mathematics, in this case points, are set into correspondence with real or idealized physical things. The mathematical characterlessness of those objects, which supports the analogy with fictional characters, also supports my contention that it is the relations among them that are being compared with the relations among the physical objects in such applications. For instance, the three bodies are not really compared with Euclidean points, but rather it is the distances among Euclidean points that are compared to the distances among the three bodies. It is when so applied that mathematics becomes true in the sense in which 'the sky is blue' is true, not by deduction from premises but by correspondence or however factual truth is understood. Since, however, what can successfully be applied is not necessarily non-fiction, this oft-realized possibility speaks weakly to the validation of the mathematics applied, however useful it is in error detection.

What we usually mean by 'true' in mathematics is the weak sense that usually coincides with 'validly deducible'—analogous to 'true in the story'.[5] As Mark Balaguer [2001] has elaborated as his *intention-based partial objectivism*, what we mean is more like following 'from the notions, conceptions, intuitions, and so on that we have in connection with the given branch of mathematics' (p. 90). Before we axiomatize arithmetic, we know how we want the positive integers to behave. If the axiomatization gives us that behaviour and only that behaviour, then we are happy with it, both its axioms and our deductions from them. If we had no idea beforehand what we wanted, then we should not know for some time whether our axioms were satisfactory even if we trusted our deductions, as happened with non-Euclidean geometry.[6]

As in a story, there are questions that can be asked in mathematics to which no answer can be given because there is no text in which to look them up (Erdös's 'Book') and we cannot deduce them either. The deductive systems where questions always have answers are too simple for mathematics. Besides the value represented by validity there is another value revealed by proof, significance. A really good idea in mathematics, like Descartes' representation of loci by equations, is not cashed out by proving it but by proving things *with it* (Rav [1999]); it has a revelatory power that the best stories have in their different way.

3. CONSEQUENCES

What is one to make of all this? Well, if you are like most that encounter this comparison, which is not the philosophers' comparison of mathematics with fiction, you will already have thought of a further comparison that I have not mentioned if only to frame a question. If you have encountered it before, then you can hardly expect to have this newcomer's reaction. Beyond extension, which seems possible endlessly, what is one to make of it? The question is rhetorical; I am going to say what I have made of it.

5 *Cf.* Woods [2002].

6 The notion of truth appropriate to fiction is 'getting it right against the background of human life', as Dan Isaacson puts it [1994]. This is quite different from the notion of truth appropriate to the *discussion* of fiction, which is based more internally on what the story says.

But first, I want to insert a disclaimer. I do not say and do not believe that the theorem-proof genre is narrative. I have twice come across the allegation that mathematics is one of the 'grand narratives', overarching explanatory schemes that postmodernists are supposed to identify and reject. On both of these occasions I challenged the writer to give me a reference of anyone that thought this; they were silent.

It seems to me that the analogy, such as it is, between proof, the most characteristic literary genre of mathematics, and narrative, a most basic mode of human communication, indicates something about proof, helps us to understand something about proof, that is not otherwise as obvious. First let me make the point that narrative is older and more basic than proof. There happened to be beside me as I wrote these sentences on July 16 a review of a book, *The Story Species: Our Life-Literature Connection*, by Joseph Gold, about whom the review said nothing. The book, however, is about 'our connection to language and to the stories that form our view of the world and our identity', quoting the review (*The Globe and Mail*, July 13, 2002, page D7) quoting the author. Mr Gold is far from alone in thinking that stories are crucial in showing us who we are, whether as species, nation, class, family, or individual. This is true of a mixture of stories, both true and fictional; we often do not know and do not care which of these a specific story is. Much of our personal identity is formed before we have a clear idea of the distinction. One might wonder when in the evolution of our civilization the distinction was first fully appreciated; is it yet? One reason why I am suspicious of comparing mathematics with fiction is that fiction is ill-defined. There seems to be little that is characteristic of fiction as distinct from other sorts of narrative. This is emphasized by the preference of some founders of extensive prose fiction in English for pretending to tell true stories; *Robinson Crusoe* springs to mind as an almost successful deception. So proof was invented against a background of narrative-based culture. But proofs are obviously distinct from stories; I have never seen them confused.[7] The idea that mathematics is fiction, while in my opinion *silly*, is not that silly. This is perhaps because in mathematics we have our own mode of storytelling; algorithms were used in arithmetical work for millennia before proofs seem to have appeared, whether among speakers of Sanskrit or Greek. So actual mathematical storytelling is readily available, but I have not yet seen anything but its existence (socio-cultural?) to comment upon. So, with a knowledge of algorithms, one sets out to produce a proof.

[7] I had previously thought that Jerome Bruner's chapter of [1986] distinguishing argument from narrative was unnecessary, but since the conference have seen—following a quotation from that very book—what appears to be the claim that 'these two narrative forms are not discrete' (Burton [1999]).

Obviously one will not produce an algorithm instead.[8] That is plainly the *other* characteristic literary genre of mathematics. The closest formal connection that I can see between stories and proofs is the way we indicate that one situation follows logically from another. Following here is a metaphor, and one that is further reflected in the locution 'if ... then ...', where 'then' does not indicate anything temporal as it would in an algorithm or story. At a higher level than the formal, we have what seems to me the most important similarity between a story and a proof, namely, that both begin with a situation as described and draw out what follows, where follows is merely analogous in the two situations, down to an arbitrarily chosen final situation.

What the analogy between stories and proofs does for me is support my contention that mathematics is not about mathematical objects but about the relations that hold among them and which they, like fictional characters, are invented to support, to allow us to speak in a way that is interesting because it says the things we are interested in saying. This is not the time to name the long succession of mathematicians that have said that relations rather than the relata are what mathematics is about. One example, from a forthcoming paper by Elizabeth Cooke [2003] on C. S. Peirce, is her interpretation of him as writing 'that mathematical truths are real in the sense that they refer to real relations, though the objects are only supposed' (p. 169). This claim would be open to disproof by counterexample, or perhaps, exceptions could be made. I confess that the mathematical object that may be an exception is the empty set, nothing at all has an ontological status that I have to admit is special.

It is perhaps of a little interest how the avalanche of comparison with fiction got started. Before Stephan Körner's comparison along these lines, I think that no one had simply compared mathematics and fiction; in fact *he* did not really. A prior example of what was called a mathematical fiction by J. E. Boodin in 1943 was the Pythagoreans' compounding of 'the world from numbers' (pp. 709 f.); I do not see in any of Boodin's examples of his mathematical fictions any comparison of mathematics and fiction. And what Körner said in 1966 [1967] was that there was no more ontological commitment to mathematics than to the world of a novel. It is some distance from this to the much better-known statements of Hartry Field that mathematics is like fiction in the sense of being empty of reference. I think Körner was indicating rather *independence* of reference. The comparison has now deteriorated to the level of a cliché, with James Robert Brown saying that Hilbert's attribution of 'existence' to whatever is consistent is

8 I leave aside the possible view of proofs as algorithms to produce conviction (proofs by induction being perhaps most amenable to this treatment).

'innocuous; it's a kind of fictional existence' ([1999], p. 100). I think Hilbert might have been shocked.

If one goes beyond Körner's point to compare narrative in general and proof as I have done for you, one *can* imagine a completely made-up story and a mathematical proof as extremes of a spectrum, with freedom increasing in one direction and logic increasing in the other. Between mathematics and fantasy there would appear both historical writing and legal evidence, each of them occupying a substantial range of the spectrum.[9] While there may indeed be a sense in which these forms of discourse are ranged along a spectrum, I hesitate to regard the directions as being happily determined by increases in logic and freedom. One has as much freedom in constructing logical argument as in telling a tall tale and possibly more, and if one's legal evidence does not hang together in a logical way, then it is going to be discarded. So the characteristics that appear along the spectrum are not so simple. Hayden White uses 'discourse' as a technical term for the range from pure fiction to logical demonstration, not inclusive, and gives it an analysis in terms of tropes in his [1978].[10] His chief concern is history, which is more closely comparable with mathematics than is fiction (and therefore a more interesting contrast) on account of history's aim to get it right in a stronger sense than that of fiction despite using the same means.

I do not myself think that the similarities, such as they are, between narrative and proof are directly explanatory of anything but rather constitute data to be explained (Thomas [2000], [2002]). There appear to be similarities of no philosophical interest, for example, the linearization that necessitates lemmas and flashbacks. In neither a proof nor a story does it make good rhetorical sense to arrange every statement in the earliest position in which its terms make sense to the reader. One's mind needs a more convenient structure than that if one is going to hold the whole together in order to understand it. Not only is linearization a necessity because we can read only one thing at a time, but the order that linearization produces is for our psychological convenience.

9 Reading Burton [1999] after the conference raised for me the questions where on this spectrum to locate a failed deduction, that it, something intended to be a logical deduction at one extreme of the spectrum that contains a flaw invalidating it, and a true fiction, that is, something purporting to be a work of fiction (any resemblance to persons living or dead is purely coincidental) but factual in content. I draw attention to the further questions: would the place on the spectrum be the same if the purported proof were fraudulent rather than mistaken, and—a question that has been discussed—what if a work of fiction is *accidentally* factual.

10 I am grateful to Sandy Berkovski at the conference for pointing out to me this fascinating book. See the first five chapters.

4. PHILOSOPHICAL INTEREST

There are, on the other hand, a couple of similarities of philosophical interest. One is the purpose for which the comparison was first introduced by Körner, to illustrate the lack of ontological commitment required in doing mathematics. There is, in reading a story that is fiction, a certain level of make-believe that one engages in and in which the author also engaged. One has to get somewhat into the picture, engage with the world of the story, though only as a spectator. This level of pretence is shallow. Perhaps even shallower is the one that Mark Crimmins has attributed [1998] to everyone when one makes a statement like 'the morning star is the evening star'. In order for the identity statement to be informative, one pretends, he says, that they are not the same thing if only briefly. It is a semantic necessity according to him. He also uses such shallow pretence to make sense of statements like 'Sam is as big a hypocrite as Uriah Heep'. In order for such a statement to make sense, one needs to pretend briefly that Uriah Heep is real so that Sam can be compared meaningfully with him. Once one has made the comparison and assigned a degree of hypocrisy to Sam, one ceases to pretend that Heep is real. Whether one chooses to express what one does in allowing ordinary semantics to apply to mathematical objects by the pretence of Mark Crimmins, which is the pretence of Kendall Walton[11] [1990], or by some other device, one does need to do something. One *may* be convinced that all mathematical objects are real. But for the large number of persons that are only lukewarm platonists or are not platonists at all—whether by leaving existence as an open question or by rejecting the existence question or by thinking mathematical objects do not exist, one needs something to justify talking about them in the usual way. I am rather taken with Crimmins's approach because it is uniform over a large field of problematic or fictitious entities and seems to work. I mention this because lack of ontological commitment is regarded as a scandal by some philosophers and some mathematicians—or even as impossible. I think that the commitment that one needs to mathematical objects is one of interest not of existence. I have mentioned Balaguer and Crimmins, but there are others that are advocating a lack of ontological commitment, Steve Yablo at MIT, and Mary Leng and Jody Azzouni at this conference (and in *Philosophia Mathematica*) come to mind.

Another similarity of philosophical interest is the way that the characters in fiction are invented to perform the function of having certain relations to other characters in the story. King Lear is constituted by the relations that he has in the play to the other characters. He does not have parents or even a

[11] Walton regards stories as prescribing certain imaginings in a game of make-believe.

wife, just daughters; he has those connections given him by the playwright and those we are meant to presuppose in order to make sense of the play. Such objects are incomplete. Likewise mathematical objects can be regarded as having only those characteristics that are given to them. The Euclidean plane is not coloured. A regular tetrahedron is neither transparent nor opaque; it brings less even with it than Lear. And so on. Whether one wants to be a structuralist or not, only structural relations of mathematical objects count. It is not clear to me that adding structure to the objects and their relations when the objects have already been added to the relations does anything for us. To me the structure is just a different sort of reification of the relations. The objectivity of mathematics is based on real-world relations not on objects either of the real world or of an imagined one, as Peirce said. If one takes the relations to be what the mathematics is about, then the sometime problem of application of mathematics is solved. Instead of having to wonder how the facts about existent mathematical objects can be transferred to physical or other non-mathematical objects, one sees that statements made about whatever has certain relations can be applied to specific items that one assumes to have those relations. They do apply by that supposition, usually to idealizations. Another problem solved is why we have different axiom systems and therefore apparently different objects for what seem informally the same system. What is the same? The relations that the formal systems are trying to capture.

I have just spoken about one of the similarities between fiction and mathematics. But within this one, as within every similarity I have encountered, lies a difference. The difference is that the relations in mathematics are purely relations among the objects, whereas the relations among fictional characters have their roots in ourselves. As the well-known Canadian novelist Michael Ondaatje was quoted in a magazine in September as saying, 'My novels are about jazz in New Orleans, or war in Sri Lanka, or in Italy. But they're also about me asking how would I behave in that situation? How would people I know behave?' (*Maclean's* 115, No. 36 (2002 09 09), 40–41) It is the presence of this personal aspect that gives power to fiction and its absence that gives power to mathematics.

The semantic point is perhaps more subtle than I have imagined. So let me spell it out a bit. Mark Balaguer has suggested to me that for me to say that what mathematical discourse is really about is the relations rather than the objects is to endorse non-standard semantics. Since standard semantics focuses so narrowly on objects and so little on the relations among them, I thought at first he was right, but I think I want to resist the suggestion. The semantics I require is wholly dependent on standard semantics, because what I am advocating is accomplished by using standard semantics for statements about mathematical objects, perhaps pretending they exist if you don't believe they do. These objects, I contend, are like pronouns that have or need no referents. Moreover, standard interpretations of 'the cat is on the mat' do not ignore altogether the relative positions in spite of concentrating on the cat and the mat. So standard semantics does take the necessary account of the relations, and I am certainly not suggesting any modification of what a mathematical statement *means*. I do not mean to suggest that 'a point is on a line' is any less precise than 'the cat is on the mat'; in particular it is not metaphorical. What I am after is that the focus of interest in the document as a whole is the relations and that the grounding in reality is in the relations; it is important that mathematical discourse be grounded. So the idea is to use standard semantics for the pronominal symbols of mathematics to avoid reifying the relations, which standard semantics would do if the relations were considered as things rather than as relating. All this is parallel to standard semantics for the pretence-based interpretation of fiction as about characters. There too there are relations, but relating rather than reified as nouns. The reference, such as it is, to relations in both cases is totally dependent on what is said of the things related, and no change in meaning is involved. In the case of fiction, one can think of either standard semantics applied to a text, making *Oliver Twist* be about a boy or non-standard semantics, making it be about the treatment of orphans. It is hardly surprising that one could use different semantics for interpretations so different, but the non-standard semantics is so heavily dependent on the standard and so small an extension of it that I am inclined to try to avoid it. If the matter is left to the interpretation of statements that plainly are, using standard semantics, about the characters and places and things of the fiction, then the relation interpretation becomes a matter of pragmatics, where it is perhaps best to leave it. This is a matter for the literary types in any case. Moreover, in the case of fiction, very little is said about the relations; they are mostly just exemplified. And we are not free to ignore the orphans. Harold Hodes calls what we do encoding the higher-level stuff we want to say in terms of the lower-level things we talk about. It comes down to whether one needs non-standard semantics to cope with the word whatever. One of the few persons to address this issue is Bas van Fraassen, who has defended realist semantics for mathematics, despite not subscribing to a realist ontology. He pointed out in his review [1975] of Putnam's *Philosophy of Logic* [1971] that the inferential structure we use relies on standard semantics. '. . . why provide another one? . . . The realist has clearly described the picture that bewitches us—the picture that guides inference. Let us just add that this is a matter of make-believe. We speak 'about' mathematical entities as if we are speaking about real things—it matters not at all whether they *are* real.' ([1975], p. 742)

There is a big difference between what is accomplished in the fictional and mathematical discourses. The mathematical, because it deals with whatevers, achieves generality directly, but fiction, because it deals with examplars, particulars, achieves generality—when it does so—only by the figure of speech by which the particular is used for the general, synecdoche. What I advocate here is that the relational focus of mathematics suggests solutions of a number of philosophical problems, whereas a narrow interpretation of standard semantics, which leads if one is not careful to thinking platonistically, may at best be psychologically helpful for actual work in mathematics. Of that, I am not convinced. It does seem to be relatively harmless provided that it does not harden from psychological help to metaphysical hindrance.

Let me try to be clearer about what it is I am trying to say by putting it this way. I am saying that the interpretation of *Oliver Twist* that makes it about the character Oliver rather than about child welfare, like the interpretation of mathematics as being about mathematical objects rather than about whatever has the relations they instantiate, is superficial. To say that they are both superficial is not to say that they are wrong. Solids *have* surfaces. To speak of their surfaces is perfectly legitimate and may be what you want to do. But if you are interested in the solids, then the superficial is only a means to gaining access to the whole, which includes the interior. The interior has more dimensions than the surface, and it may be that there is something to be gained by paying attention to the multidimensional interior instead of assuming that the surface is all that there is. In suggesting what can be thought of as a slight extension of standard semantics, I am by no means suggesting something independent of standard semantics but rather what should be a standard extension since it is needed for serious talk about fiction, science, mathematics, and all thought about the future.

I want to illustrate the way that ordinary semantics needs to be preserved by showing how it is similarly necessary in allegory, where the real focus of attention is relations that are not spoken of but are referred to only by analogy. Consider the allegory that is famous in post-Christian countries, sufficiently famous that in Britain the logo of the Bible Society, which distributes Christian scriptures in many languages all over the world, was until recently a man sowing grain by hand. In the three authoritative records of the doings and sayings of Jesus, he tells this story: A sower went out to sow: and as he sowed, some seed fell by the way side, and the fowls of the air came and devoured it. And some fell on stony ground, where it had not much earth; and as soon as it sprang up, because it had no depth of earth, it withered away for lack of moisture. And some fell among thorns, and the thorns grew up, and choked it. And other fell on good ground, and sprang up and bare fruit an hundredfold. This little story, which is called a parabolé (the word that gives us parable in English) in the text, is then in each of the texts immediately given an allegorical interpretation. The story is about preaching. The fowls of the air are spiritual distractions; stony ground is interpreted as hearers that are initially attracted but lose interest; the thorns are the cares of the world, less spiritual distractions; but the good ground is receptive hearers and justifies the whole process. My point in showing off this tiny allegory is to point out that allegorical interpretation is not a new sort of semantics but something that is done to the meaning of the text rather than to the text itself. One needs the initial semantic decoding of the text to be already accomplished before the allegorical interpretation can begin because the raw material of the allegory is what the story is literally about and not the story's words themselves.

One of the things I have found most fascinating about this analogy is that, as I have previously said, within each similarity is a difference. Application of mathematics is like allegorical interpretation. But. Allegory is applied to narrative, and so the structure that it transfers is narrative structure. Theorems, on the other hand, have logical structure, and when a theorem or body of theorems is applied the logical structure transfers to the target domain. I have little of use for allegorical interpretation, because there is no force to it; the correspondences are just brute facts based on brute stipulations. But applied mathematics gives correspondences that have logical necessity because of the transfer of the structure, which is not merely narrative but is logical.

I do not want to appear to think that narrative is the only thing we can profitably compare mathematics to. There is also science, which is a closer comparison. I end by saying a little about that. I view mathematics as similar to science and to be understandable by those that have some understanding of science on that basis because science in the post-Galilean sense imitates mathematics in this specific way. It ceased to concern itself with what things are and has concentrated on how they are related. No chemist would take seriously a discussion of the essence of antimony; that went out with alchemy. Once one is dealing with relations of physical sorts among physical things, one is doing something rather like mathematics; it is no wonder that it took some time after physics took this turn to distinguish pure geometry from the study of space. I think that an important step in that distinction was acquiring the freedom from ontological commitment. So in mathematics we do the same sorts of thing that theoretical physicists do, except that the relations that we study are among whatever rather than among physical things, even idealized physical things. I do not know enough about philosophy of science to know whether taking this view, which I know I have not invented, will help philosophy of mathematics by allowing it to imitate philosophy of science. I mention the possibility because it does seem to me that philosophy of physics is more closely focussed on actual physics than philosophy of mathematics is focused on actual mathematics.

In the month following the conference, John Woods published the first serious look ([2002], Chapter 6) at the logic of fiction for years. It is a look at only the logic of fiction, so that it shares with Zalta [1983] the limitation that it considers only what one might call the world of the work rather than the narrative itself. It also gives an unsympathetic caricature of Walton's aesthetic theory in discussing it. It is, however, an important document, not least for describing a view that some may have found tempting:

> Fictionalism: Since fictionalism with regard to a theory T proposes that T's truths are secured and validated by analogy with how the truths of fiction themselves are secured and validated, fictionalism with respect to T cannot be an adequate semantic theory of T unless the theory of fiction borrowed by T's fictionalism is in its own right a semantically adequate theory of fiction. (p. 196)

On serious consideration of this view, Woods concludes that rather than modelling abstract sciences on fiction (p. 222), it should be 'the other way round' (p. 223). For my purposes, I have not needed a logic of fiction nor been tempted to model mathematics on fiction, but I certainly agree that fiction is the tougher nut to crack.

REFERENCES

Balaguer, Mark [2001]: 'A theory of mathematical correctness and mathematical truth', *Pacific Philosophical Quarterly* 82, 87–114.

Boodin, J. E. [1943]: 'Fictions in Science and Philosophy I' and '. . . II', *J. Phil.* 40, 673–682 and 701–716.

Brown, J. R. [1999]: *Philosophy of mathematics*. London: Routledge.

Burton, Leone [1999]: 'The implications of a narrative approach to the learning of mathematics', in Leone Burton, ed. *Learning mathematics: From hierarchies to networks*. London and Washington: Falmer Press, pp. 21–35.

Cooke, Elizabeth F. [2003]: 'Peirce, Fallibilism, and the Science of Mathematics', *Philos. Math.* (3) 11, 158–175.

Crimmins, Mark [1998], 'Hesperus and Phosphorus: Sense, Pretense, and Reference', *Philos. Rev.* 107, 1–47.

Eco, Umberto [1979]: *The role of the reader: Explorations in the semiotics of texts*. Bloomington, Indiana: Indiana University Press.

Field, Hartry [1980]: *Science without numbers: A defence of nominalism*. Princeton: Princeton University Press.

----------- [1989]: *Realism, mathematics and modality*. Oxford: Blackwell.

----------- [1990]: 'Mathematics without truth (a reply to Maddy)', *Pacific Philosophical Quarterly* 71, 206–222.

Isaacson, Daniel [1994]: 'Mathematical intuition and objectivity', in Alexander George, ed. *Mathematics and mind.* New York: Oxford University Press, pp. 118–140.

Körner, Stephan [1967]: 'On the relevance of post-Gödelian mathematics to philosophy', in I. Lakatos, ed. *Problems in the Philosophy of Mathematics.* Amsterdam: North-Holland, pp. 118–132, with discussion by Gert H. Müller and Y. Bar-Hillel and reply, pp. 133–137.

Paulos, John Allen [1998]: *Once upon a number: The hidden mathematical logic of stories.* New York: Basic Books.

Putnam, Hilary [1971]: Philosophy of Logic. New York: Harper and Row. Reprinted in his [1975], pp. 323–357)

---------------- [1975]: *Philosophical Papers. Vol. 1.* Cambridge: Cambridge University Press.

Morton, Adam [1996]: 'Mathematics as Language' in A. Morton and S. P. Stich, eds. *Benacerraf and his Critics.* Oxford: Blackwell, pp. 213–227.

Rav, Yehuda [1999]: 'Why Do We Prove Theorems?', *Philos. Math.* (3) 7, 5–41.

Rotman, Brian [1993]: *Ad Infinitum: The Ghost in Turing's Machine.* Stanford: Stanford University Press.

Thomas, R. S. D. [2000]: 'Mathematics and Fiction I: Identification', *Logique et Analyse*, 43, 301-340.

---------------- [2002]: 'Mathematics and Fiction II: Analogy, *Logique et Analyse*, 45, 185-228.

---------------- [2002]: 'Mathematics and Narrative', *The Mathematical Intelligencer* 24, No. 3, 43–46.

Turner, Mark [1996]: *The literary mind.* New York: Oxford University Press.

van Fraassen, Bas [1975]: review of H. Putnam's Philosophy of Logic. *Canadian J. Phil.* 4, 731–743.

van Inwagen, Peter [1977]: 'Creatures of fiction', *American Philosophical Quarterly* 14, 299–308.

Walton, Kendall L. [1990]: *Mimesis as Make-Believe: On the Foundations of the Representational Arts.* Cambridge, Mass.: Harvard University Press.

White, Hayden [1978]: *Tropics of Discourse: Essays in Cultural Criticism.* Baltimore and London: Johns Hopkins University Press.

Woods, John [2002]. *Paradox and Paraconsistency: Conflict resolution in the abstract sciences.* Cambridge: Cambridge University Press.

Zalta, Ed [1983]: *Abstract Objects: An Introduction to Axiomatic Metaphysics.* Dordrecht: Reidel.

Chapter 4

THEORY OF MIND, SOCIAL SCIENCE, AND MATHEMATICAL PRACTICE

Sal Restivo
Rensselaer Polytechnic Institute

Abstract: The very idea of "mathematical practice" implies, beyond the transparent social turn in philosophy, pedagogy, and didactics of mathematics, a theory of mind. Theories of mind may be the informal folk theories of our everyday lives or the more formal theories of professional students of mind. My conjecture is that folk theories of mind are at present still more influential in the work of students of mathematics and the mathematics classroom than are professional theories. The problem is that whichever theory prevails in any given setting or study, or for any given researcher, it is more likely than not to locate the mind in the brain and in the person. So one question I want to pose is: which theory or theories of mind are built into our theories of and approaches to mathematical practice? If turning our attention to mathematical practice as opposed to focusing on questions of foundations is a turn to the social, perhaps we should be alert to the possibility of a social turn in our theories of mind.

Key words: Mathematics, sociology, mind, mathematics education

1. THEORIES OF MIND

Traditionally, theories of mind, primarily coming out of philosophy and psychology, have been asocial. These theories include Hegelian mentalism, idealism, materialism, dualism, various forms of monism, and variations on these themes including Cartesian, bundle, interactionist, parallelist, behaviorist, logical behaviorist, functionalist, phenomenological, central state or identity theories, and various attribute theories (Armstrong, 1968; Priest, 1991). One of the most intriguing aspects of how philosophers and psychologists think about minds is the effort to explore the nature of the human mind by imagining about brains in vats, armadillo minds, thinking

B. Van Kerkhove and J.P. van Bendegem (eds.), Perspectives on Mathematical Practices, 61–79.

bats, and Martian brains. No wonder we can't find social human beings anywhere in these theories!

These traditional and prevailing approaches to mind and mentality in general center on the brain. Mentality is viewed as either caused by or identical with brain processes. Given this perspective, John Searle (1984:18) could argue that "Pains and all other mental phenomena are just features of the brain (and perhaps the rest of the central nervous system)." But Durkheim's analysis of different degrees of social solidarity and the social construction of individuality suggests a culturological conjecture on pain: the extent to which a person feels pain depends in part on the kind of culture s/he is a product of, and in particular the nature and levels of social solidarity in the social groups s/he belongs to. Furthermore, the symbolism of the pain experience in its cultural context is also a determinant of felt pain. "Pain" has a context of use, a grammar. Such a conjecture was indeed already formulated by Nietzsche (1956/1887: 199-200) in *The Genealogy of Morals.* Wittgenstein's (1953) writings on pain in his *Philosophical Investigations* provide additional ingredients for a social theory of mind based on the role of language in our pain narratives. But Searle, while he invokes the social, does not know how to mobilize it theoretically, and so argues that consciousness is caused by brain processes. We will see as we proceed why this claim that has seemed so reasonable for so long must be reconsidered in light of what we know about the relationship between social life and consciousness, and what we are learning about social life and the brain.

Cognitive psychologists tend to view the mind as a set of mental representations. These representations are then posited to be causes behind an individual's ability to "plan, remember and respond flexibly to the environment" (Byrne, 1991: 46). Cognitivists also have a tendency to equate cognition and consciousness. But Nietzsche long ago had the insight that consciousness is a social phenomenon. He was one of a number of classical social theorists who had pioneering insights into the social nature of mentality.

We can approach the history of discourse on mind in terms of (1) the conflict between rationalists (intellectual descendants of Descartes and Leibniz) and empiricists (followers of Locke, Berkeley, and Hume); (2) the behaviorist challenge to the radical empiricists by Watson and others, and the challenge in turn to the behaviorists by the ethologists (Lorenz, Tinbergen, and von Frisch); and (3) the Kantian counterpoint to empiricism, represented in our own time, for example, by Jerry Fodor's (1983) conception of the mind as an entity possessing organizing capacities and an innate "language of thought."

Why is it we "locate" mind, thinking, and consciousness "inside" heads? Certainly in the West, mentalities and the emotions have been associated with the brain and the heart since at least the time of the ancient Greeks. More recently, localizationalist physicians and neuroscientists have reinforced the idea that mentalities are "in the head" (Star, 1989). On the other hand, in sociological perspective, mentalities are not produced out of or in states of consciousness; they are not products, certainly not simple products, of the evolution of the brain and brain states. Rather, they are by-products or correlates of social interactions and social situations. This implies that the "unconscious" and the "subconscious" are misnomers for the generative power of social life for our mentalities – and our emotions. There is no more an unconscious than there is a God, but there are cultural mechanisms for translation and transference that point us to referents that do not exist. The thesis here is that social activities are translated into primitive thought "acts," and must meet some filter test in order to pass through into our awareness (cf. Wertsch, 1991: 26-27; and see Vygotsky, 1978, 1986; and Bakhtin, 1981, 1986). Vygotsky and Bakhtin should be considered independent inventors of the modern social theory of mind alongside their contemporary, G.H. Mead. Wertsch (1991: 14) stresses that mind is mediated action, and that the resources or devices of mediation are semiotic. Mind, he argues, is socially distributed mediated action.

2. GETTING TO THE BEGINNING OF OUR STORY

In 1939, C. Wright Mills (1963) argued that the sociological materials relevant to an understanding of mind had not been exploited by sociologists. Mills had in mind in particular the work of the American philosopher and social theorist, George Herbert Mead. Fifty years later, Randall Collins (1988) could still write that Mead's writings on the sociology of mind were underdeveloped and unexploited. And as I write these words more than a decade later, the same claim can be made. Indeed, the failure of sociologists to pick up the track of Mead's social theory of mind was underscored by the publication of a social theory of mind guided by Mead's work written not by a sociologist but by a neuroscientist/psychiatrist (Brothers, 1997).

3. RESOURCES FOR A SOCIOLOGICAL THEORY OF MIND

The basic resources I draw on for constructing a sociological theory of mind include but are not limited to the following: the concepts of *collective representations* (Durkheim) and *generalized other* (Mead); Goffman's

(1974) *theory of frames* (cf. Wertsch, 1991 on *recontextualization*); the literature on *culture and thought* (Levy-Bruhl, 1985/1926; Levi-Strauss, 1966; Goody, 1977; Cole and Means, 1981); studies of the evolution of *human language and its social context* (Caporael et al., 1989), and studies *on the relationship between social relations and rule-governed systems such as language* (Caporael, 1990: 10-11). Researchers in artificial intelligence have been increasingly incorporating into their work the idea that *AI machines have to be programmed with "cultures"* (e.g., Normal and Rumelhart, 1975; Keesing, 1987: 381). By the 1990s, these and related ideas had coalesced into efforts to build affective computers and social robots (e.g., Picard, 1997; Breazeal, 2002; Restivo, 2001).

It is also important to register in these early moments of this effort in theory construction that sociology has something to say about the brain. Clifford Geertz (1973: 76) has pointed out that the brain is "thoroughly dependent upon cultural resources for its very operation; and those resources are, consequently, not adjuncts to but constituents of, mental activity…" Indeed, DeVore (Geertz, 1973: 68) has argued that primates literally have "social brains." The evidence for this conjecture in humans has been accumulating in recent years along with a breakdown of the brain/mind/body divisions (e.g., Brothers, 1997; Pert, 1997).

Our understanding of mentalities has been obstructed by some deeply ingrained assumptions about human beings. One is that affect and cognition are separate and separated phenomena. This division is breaking down (e.g., Zajonc, 1980, 1984; Gordon, 1985; Damasio, 1994; and Pert, 1997), and will have to be eliminated as part of the process of constructing a sociology of mind.

Another assumption is that learning and cognition can be decontextualized. I argue with other social scientists, by contrast, that learning and cognition are linked to specific settings and contexts, that is, they are indexical. Their long-term efficacies are in fact dependent on contextual recurrence, contextual continuity, and recursive contextualizing. The latter process helps explain the process of generalization without recourse to epistemological mysteries or philosophical conundrums. We live our lives by moving from home or school to home or school, from our home to our neighbor's home, from the schools we attended to the schools our children attend. Contexts repeat, imitate, suggest, overlap, impose and re-impose themselves, shadow and mirror each other, and are linked through simple and complex feedback loops. This is the structural and informational basis for the continuities in our sense of self, our memories, our thoughts. Many of the mysteries of the paranormal and our everyday experiences of déjà vu can be explained by attending to these features of context.

According to Astington (1996: 184), Gopnik (1996) claims there are only "three games in town" when it comes to theory of mind: theory-theory, simulation theory, and modularity theory. But Gopnik (1996: 169, 182) distinguishes "theory-formation theory" from "theory-theory." Nonetheless, the psychologism in these theories fits the individualist bias we find in work ranging from research on children's' theories of mind to social robotics. There is another game in town, however, and it goes with the sociological resources I have sketched. The alternative to children deriving their theories of mind from their direct experiences of such states, of developing such theories the ways scientists supposedly derive their theories, or of giving rise to them innately as they mature is an enculturation theory. The prevailing theories of theory of mind emphasize development within the individual. From a sociological or anthropological perspective, theory of mind and mind itself are cultural inventions (Astington, 1996: 188). Social construction of mind has not been ignored, but it has not been as centrally represented in either mind studies or social robotics. The reason is a problem in the sociology of knowledge. It may be, for instance, that it is easier to link psychology and engineering because psychology appeals to the illusion or fallacy of introspective transparency. The problem with sociology is that while it holds unparalleled promise for social and sociable robots engineering, it is by comparison with the psychological sciences introspectively counter-intuitive and technologically sterile. These are not failures of sociology but rather failures of the sociological imagination in robot science and engineering. Similar problems accrue to educational theories in mathematics to the extent that they are grounded in traditional psychology and philosophy.

4. THE SOCIAL MIND

The sociology of mind and thinking has a long and distinguished pedigree, yet it has until recently been virtually invisible in contemporary theories of mind (Valsiner and van der Veer, 2000). A renewed interest in mind, brain, consciousness and thinking (along with the new life evident in the search for God – the two quests are indeed related in sociology's program for the rejection of transcendence) is evident in the steady stream of books, articles, lectures, news stories, and television programs crossing today's intellectual landscapes. One of the main features of this literature is that one can see some evidence of a sociological orientation emerging, albeit timidly and fearfully, out of the shadows.

An archaeology of these developments would reveal a "journey to the social" across the entire landscape of intellectual labor. Virtually without exception, those who undertake this journey are not sociologists or anthropologists (or more generally, social scientists) and so they stop short of their mark or otherwise abort the trip. This is, indeed, a much more treacherous journey than the Westerners' journeys to the east which have captivated (and captured) so many Western seekers. But the very fact of the journey to the social reveals the emergence of a new discursive formation, a new episteme.

This episteme is new in the sense of a birth or an originating activity, but absolutely new in the scope of its impact. Beginning in the 1840s, the West entered the Age of the Social, an era of worldview changes that will carry well into the 21st century and likely beyond before it begins to embody itself in the everyday ecologies and technologies of mind in new global configurations. In this process, what was western and European about the social will get permeated and transformed into a worldview that is less ethnocentric.

Most immediately, nothing captures the spirit of this renewal better than philosopher John Searle's *The Rediscovery of the Mind.* Searle (1992: 128), for example, writes:

> I am convinced that the category of "other people" plays a special role in the structure of our conscious experiences, a role unlike that of objects and states of affairs...But I do not yet know how to demonstrate these claims, nor how to analyze the structure of the social element in individual consciousness.

And the neuroscientist Antonio Damasio (1994: 260) writes:

> To understand in a satisfactory manner the brain that fabricates human mind and human behavior, it is necessary to take into account its social and cultural content. And that makes the endeavor truly daunting.

To give one more example, consider the following remarks by Stan Franklin ((1995: 10) at the beginning of his tour of mind studies. Franklin is a mathematician and computer scientist:

> Let's not leave our discussion…without pointing out its major deficiency. There's no mention of culture. How can one hope to understand mind while ignoring the cultural factors that influence it so profoundly? I certainly have no such hope. I clearly recognize that the study of culture is indispensable to an understanding of mind. I simply don't know how to gently include culture….Perhaps anthropology and sociology should share a corner with cognitive psychology.

And even in artificial intelligence research, projects from Rodney Brooks' COG (the baby robot) to the view of mentality as "physically and environmentally embedded" (Torrance, 1994), and the idea of cognition as embodied action (Varela, Thompson, and Rosch, 1991), paths are being opened for social and cultural studies of mentality.

5. THE SOCIOLOGY OF MIND

The idea that the mind is a social construction is crucial to reforming our understanding of mathematics education in the light of the sociological perspective. I come to the sociology of mind by way of the sociology of science, mathematics, and knowledge. In particular, I have been concerned over a major part of my research career with bringing mathematics down to earth. To bring mathematics out of the Platonic clouds, out of transcendental realms, is equivalent to negating the idea of "pure mind."

When, and to the extent that, mathematics becomes a functionally differentiated, institutionally autonomous social activity in any given social formation, it will begin to generate mathematics out of mathematics. The vulgar notion that "mathematics causes mathematics" (pure mathematics) arises out of a failure to (and to be able to) recognize that in a generationally extended mathematical community (or social network of mathematicians), mathematicians use the results of earlier generations of mathematical workers and mathematicians as the (material) resources for their mathematical labors. Systematization, rationalization, generalization, and abstraction in mathematics are dependent on organizing mathematical workers in a certain way. In general, this means specialized networks and sustained generational continuity. The widely recognized significance of iteration as a factor in mathematical development and creativity is dependent on the social iteration produced by generational continuity.

For centuries, it has seemed obvious that the study of mind should be under the jurisdiction of philosophers and psychologists (in their pre-modern as well as modern guises). As the matrix of mind studies became increasingly interdisciplinary in the latter part of the last century, sociology and anthropology were notably left out in the cold. It may be that these are the only modes of inquiry that have any hope of making sense out of the chaos of claims about mind, consciousness, and even God and soul coming out of contemporary physics, astronomy, biology, artificial intelligence and the neurosciences.

In 1943, Warren McCulloch and Walter Pitts helped set the agenda for an immanentist approach to mind. They claimed that "LOGIC is the proper discipline with which to understand the brain and mental activity. The brain embodies logical principles in its neurons." Durkheim had already rejected immanence along with transcendence in *The Elementary Forms of (the)*

Religious Life. That is, he rejected in the first instance the notion that ideas such as Aristotle's categories and Kant's categorical imperatives are either (a) logically prior to experience, immanent in the human mind, or a priori; or (b) crafted by individuals. In the second instance, he rejected the idea that there are transcendental referents (for terms, for example, such as "soul," "God," and "heaven"). The crystallization of the rejection of immanence and transcendence is one of the great on-going achievements in the history of thought. The project arguably begins as early as Socrates. Cicero said that Socrates "called philosophy down from the sky..." A more recent example of this imperative is Dirk Struik's (1986: 280) conception of the goal of the sociology of mathematics: to haul the lofty domains of mathematics "from the Olympian heights of pure mind to the common pastures where human beings toil and sweat."

Sociologists like Randall Collins and myself begin our efforts in the sociology of mind by making a simplifying assumption - that thinking is internal conversation. This poses an immediate problem. That is, given everything I have written so far, and given Wittgenstein's writings on mind and thinking, I do not want to claim that thinking (as conversation, for example) is something that happens inside heads or brains. There are efforts abroad to develop an explanation of cognition as embodied action. A theory of embodied action that is properly sociological dissolves the inner/outer dilemma and the chicken/egg problem. The chicken point of view is that there is a world "out there" with pre-given properties. These exist before and independently of the images they cast on the cognitive system. The role of cognition is to recover the external properties appropriately and accurately (Realism). From an egg perspective, we project our own world, and "reality" is a reflection of internal cognitive laws (Idealism). But a theory of embodied action explains cognition/mentality in terms that depend on having a body with a variety of sensorimotor capacities embedded in more encompassing biological, psychological, and cultural contexts (Varela, Thompson, and Rosch, 1992). Cognition is lived; sensory and motor processes, perception, and action are not independent. This approach promises to dissolve the inner/outer dilemma, and to eliminate representational paradoxes in the theory of mind. Details on how such a perspective bears on our understanding of how we learn mathematics can be found in Stephen Lerman's (1994) work. There is already more than a hint in the embodiment imperative on how to solve the mind/body problem. We do not socialize individuals, persons, or selves as sociology textbooks and theorists are wont to claim. That claim assumes as already extant what socialization aims to create. What then is socialized (assuming that that is still an appropriate term)? Suppose that what is socialized is the brain/central nervous system, with the b/cns embedded in a social ecology, a network of social relationships and interactions? A theory constructed along some such line would eliminate the mind/body problem and make a rapprochement between neuroscience and social science a reasonable expectation.

In order to cement the relationship between embodiment and the sociology of mathematics, consider what we mean by endless counting. I am inspired here by Nietzsche, Wittgenstein, and Rotman among others. We must ask in the first place who is it that will be making the endless marks or speaking the endless numbers? It will have to be us or some surrogate. We could only relate to such a surrogate if it was like us in significant ways. If what is significant about us is our embodiment, any counting surrogate would have to be embodied. The alternative is a disembodied agent counting outside of time, space, and materiality. An embodied counter undermines Platonism and absolutism and gives us the grounds for a sociology of mathematics. Numbers then appear as things embodied humans have to make and re-make.

For the moment, I want to focus on the "phenomenology" of a certain kind of thinking experience. A sociological theory of mind must account, one way or another and sooner or later, for the experience of "inner thought." And it must do so without the assumption or claim that this experience is universal across humans and cultures.

Conversation is the prototype for a certain kind and level of thinking, the kind of thinking we, initially at least, have in mind (so to speak) when we set out to construct an artificial intelligence, develop a theory of mind, or think about our own thinking. We must learn to speak out loud before we can think "silently," "in our heads." "External" speech already contains all the crucial elements of thought: significant symbols, capacity to take the stance of one's interlocutor or listener, and ability to take the role of the other and orient to the generalized other (see the discussion below).

Internal conversations do not necessarily have the same structure as external conversations. Short cuts, shunts, and short-circuits in our thinking are possible when we (as adults) are thinking smoothly. We may know almost immediately where a thought is going and whether to pursue it or switch over to another thought-track. Because we can monitor multiple thought-tracks (the dispatcher function), we can rapidly switch between alternatives, elaborations, objections, and conclusions. Thought-tracks and trains of thought connect syntactically and pragmatically in Hesse-type networks (Hesse, 1974). And words invoke other words, ideas invoke ideas, concepts invoke concepts (because of similar meanings, sounds, and/or associations). Generally, these switches, invokings, and associations occur smoothly and without the exercise of "will;" and they can produce what I call thought cascades. If the process is disrupted in any way, however, our attention will shift, the process will slow, and we will proceed with awareness. This contributes to the illusion that we think "willfully."

If we treat thinking as internal conversation, then thinking must be constructed out of past, anticipated, and hypothetical conversations. In other words, what we think is connected to our social networks (including reference groups). Then the greater the attraction to given parts of the network, the more we will "be motivated" to think the ideas circulating in those parts.

The connections among ideas are emotional as well as associative and grammatical. Words, ideas, and images have valences. And consciousness itself is a type of emotion, attentiveness. Normally this attentiveness is very mild and attached to certain sign-relations. The level of attentiveness presumably changes as social situations (real and imagined) change. Only when the smooth and easy inference (or "next move") is blocked, or contradicted by something in the situation, does the emotion erupt into consciousness. So emotional weightings (valences) affect what a person thinks about at a particular time. These ideas are consistent with neuroscientific and sociological research that suggest the existence of a baseline emotional state.

6. THE GENERALIZED OTHER REVISITED

The generalized other is the core concept in George Herbert Mead's social theory of mind. Mead introduced the idea of the generalized other to describe that component of the self constructed out of the variety of messages we receive from the people we come into contact with. The generalized other is the source of our ability to take the roles of others, and also the source of our understanding of the "rules of the game" in everyday interaction. It is the locus of what Freud called the super-ego, which gives us "conscience." And it is the locus of what I call "moralogics." When we reason, generalized others are with us all the way, approving and/or disapproving our every move. We always reason from a standpoint. There are many standpoints, and each is guarded by a generalized other. Operating logically means operating in terms of standard and standardized critical and reasoning apparatuses. Individuals cannot be logical or illogical. They can only be in agreement or disagreement with a community of discourse, an objectivity community, a thought collective. And patrolling standpoints is therefore a moral act. If, then, reasoning is always grounded in a standpoint, there can be no General Abstract Reasoner, no eternal, universal logic. If, furthermore, patriarchy has constructed Platonism, and relativity theory, and truth-seeking Diogenes and the propagandist Goebbels, the podiums of rationality and objectivity and the arenas of emotion, then there is good reason (from a certain standpoint, now!) to conjecture that mentality or mind is "man-made." Thinking is, therefore, on these principles, gendered. Logic is the morality of the thought collective, and carries the weight of how gender and power are distributed therein.

Neither "laws of logic" nor "laws of thought" (George Boole) are intuitive, innate, or a priori. Generalized others carry socially derived logical systems that restrict, govern, filter, direct, and cue logical speech acts. Inside every word, inside every vocabulary, inside every sentence, and inside every grammar we find discourse communities; logics are language games. It follows that our thoughts, insofar as they draw on the resources of languages, are socially textured. Here Goffman's (1974) frame analysis provides another ordering apparatus. And the distinction he attends to between conversational talk and informal talk has an analogy in thought. Just as informal talk holds the individual together across parsing moments and breaks in continuity in social projects, and just as much of what we say in the presence of others is related to creating and sustaining social solidarity, so informal thought is about self-solidarity. Speaking, Goffman points out, "tends to be loosely geared to the world." Talk is looser. I conjecture that thinking is even looser, and more vulnerable to the processes Goffman calls keying and fabricating.

Now let us think again about moralogics. Mathematics communities are in part crucibles for refining the idea of God through exercises with infinity(ies). The most abstract efforts then turn out to be tied more or less explicitly to the God project. Boole's goal was to reduce "systems of problems or equations to the dominion of some central but pervading law." This is not a simple metaphor, for Boole was set on establishing the existence of God and a universal morality. So too Cantor's transfinite numbers are implicated in the search for a proof of the existence of God. I cannot pursue this further here, but see the appendix on mathematics and God in Restivo (1992).

7. WHERE IS THINKING?

The introduction to the social construction conjecture should make it easier to understand what I mean when I say that minds and thinking are social constructions. This conception carries with it the notion that thinking is a networked and dialogic process, a series of social acts rather than something that goes on inside isolated, independent heads and/or brains. This does not mean that heads and brains are dispensable, or that neuroscientists and psychologists have nothing to teach us about minds and thinking. But it is social relations that give rise to consciousness and thinking; the genesis of consciousness and thinking is in society not in the brain. Freestanding brains do not and cannot "become" conscious, and do not and cannot generate consciousness in some sort of evolutionary or developmental "brains in a vat" process. Consciousness, thought, and language cannot be explained or understood independently of the understanding that human beings are fundamentally and profoundly social. More importantly, they are profoundly rhythmic, and it is the coupling of their rhythms in social interaction that produces consciousness, emotions, and communication.

Individualized thoughts must be tied to their social bases if we are to understand their genesis and nature. Communicable thoughts are, by definition, shareable and shared (Durkheim, 1961: 485). All concepts are collective representations and collective elaborations - conceived, developed, sustained, and changed through social work in social settings. Indeed, Randall Collins (1997) has shown through detailed comparative historical studies that the configurations and developments of social networks of intellectuals cause particular ideas to come into being and develop or die out. This line of thinking leads to the conclusion that it is social worlds or communities that think and generate ideas and concepts, not individuals. Social worlds do not, of course, literally think in some superorganic sense. But individuals don't think either. Rather, individuals are vehicles for expressing the thoughts of social worlds or "thought collectives." Or, to put it another way, minds are social structures (Gumplowicz, 1905; Fleck, 1979: 39). Mentality is not a human invariant. And even vision is an activity and not a neurological event (Davidson and Noble, 1989; and see Heelan, 1983 on the social construction of perception).

In order to grasp the idea that thinking is radically social, and to keep it from slipping into some spiritual or mystical realm, or becoming an empty philosophical or theological concept, one must keep firmly focused on and fully comprehend the idea that humans are social beings and that the self is a social structure. It is also crucial that we do not project our modern post-literate experience of mentality and mind-body duality on all humans in all times and places. "Mind" is not a cultural or human universal (cf., Olson, 1986, and Davidson and Noble, 1989).

8. RITUAL AND COGNITION

Cognition arises situationally out of the natural rituals of everyday interactions and conversations. These rituals form a chain, and as we move through this chain, we come across and use more or less successively blends of cultural capital and emotional energies (Collins, 1988: 357ff.). The concept of ritual developed in the work of Emile Durkheim can be generalized and conceived as a type of framing (following Goffman, 1974). This leads to the idea that the theory of ritual can be developed in terms of the different types of framings and reframings, which constitute our movement through interaction ritual chains. From this perspective, solidarity rituals take place in a social market that is variously stratified. Language is a product of a pervasive natural ritual (words, grammatical structures, speech acts, and framings are collective representations loaded with moral significance). The ingredients of language refer outside conversations, and their sense is their symbolic connection to social solidarity and their histories in interaction ritual chains. All thoughts take place in several modalities - visual, aural, emotional, sensual - simultaneously. Indeed, it is the socially constructed, gendered, cultured body-in-society that thinks, not the individual, or the head, brain, or mind.

We are now ready to enter the world of the sociology of mathematics. But I must stress that if we enter without at least some preliminary comprehension of the ideas that self and mentality are social, the sociology of mathematics will seem like a voyage through the Looking Glass - without any of the charm of Lewis Carroll's guidance.

9. CONCLUSION: A SOCIAL CONSTRUCTIONIST PARADIGM FOR THE MATHEMATICS CLASSROOM

After all of this social theory, the mathematics educator may very well still feel at a loss as to how to translate social constructionism and its entailments into classroom practices. Whatever guidelines I have provided for such a translation may seem to be distant from the immediate classroom concerns of mathematics teachers. This is inevitable given the difficulties of coming to terms with what amounts to a new worldview. Let us see if we can come a little closer to the communication essentials of the mathematics classroom.

Given social constructionism, classroom teachers and students must learn to ask some old questions with an expectation that there will be new answers. What are numbers (and what are all the basic concepts and processes that constitute mathematics?). What is a classroom? What are teachers and students? What is learning? What is truth? What does it mean to reason? What is a proof? The trick here is to see all of these old friends as ***institutions.*** A number of mathematics educators have made significant progress in coming to terms with this perspective and making it accessible to and applicable within the working lives of mathematics teachers. (see, for example, Ernest, 1994; 1998; Burton, 1999).

Let us adopt the teacher's point of view as s/he enters the classroom armed with the tools of sociology. To begin with, the room is no longer filled with individual students housing individual, freestanding brains. It is now to be seen, to be experienced, as a collectivity, variously cultured and more or less culturally homogeneous or heterogeneous. No assumptions can be made about shared paradigms, practices, and discourses (Sfard, 1994: 248). I want to speak in terms of a focus on practice as opposed to reasoning. But this is a false dichotomy. Given the social constructionist perspective, it follows that all aspects of mentality are forms of social practice. The everyday significance of this is that mathematics teachers are not engaged in "teaching" students how to think mathematically. Rather, they are initiating students into different corners of the mathematical community, and on different levels of mathematical practice. Therefore, their first concern should not be with the ontology or epistemology of mathematical objects and ideas, but rather with the practices that give rise to and sustain those objects and ideas in the lives of students.

Sfard (1994: 270) recommends (following, for example, Leibniz's philosophy, not to mention Wittgenstein) an emphasis on using concepts in various contexts rather than trying to get students to grasp concepts immediately in the abstract. It would be fruitful if teachers in all fields recognized that learning in general emerges as we use words in different contexts and not out of brute force definitions and applications. It is not shared definitions that make communication possible but rather common practices in pursuit of shared goals.

There is the suggestion in Sfard (1994: 269-271) that mathematics students might learn more effectively by recapitulating the ways the mathematical community came to collectively grasp concepts and ideas. As a social constructionist, I would certainly advocate teaching mathematics in the context of their historical development. The historical, social, and cultural contexts cannot be separated out from the substance of mathematical objects, concepts, and ideas. As long as these dimensions of mathematics are considered, "humanistic accessories" and "nonessential distractions" from the curriculum of training, teachers will not be motivated to carry out the requisite curriculum changes. Foundational resources for such changes are readily available in the works of Restivo (1983, 1992) on the calculus, mathematical traditions across cultures, and the networks of mathematics (math worlds), MacKenzie's (1981) study of the social construction of statistics, and Bloor's (1976) pioneering program for a sociology of mathematics.

Collaborations between social scientists and mathematics teachers would be useful. But this will be counterproductive unless the social scientists are mathematically literate and the mathematicians are sociologically literate. It is still the case that sociological literacy is much more rare than mathematical literacy. Sociological literacy depends on a worldview free of certain fallacies. I conclude with a list of those fallacies. These fallacies are the raw materials for a set of theorems about our nature as social beings.

- The Transcendental Fallacy (the theologian's fallacy) is that there is a world or that there are worlds beyond our own – transcendental worlds, supernatural worlds, worlds of souls, spirits and ghosts, gods, devils, and angels, heavens and hells. There are no such worlds. They are symbolic of social categories and classifications in our earthly societies and cultures. There is nothing beyond our material, organic, and social world. Death is final; there is no soul, there is no life after death. It is also possible that the so-called "many worlds interpretation" in quantum mechanics is contaminated by this fallacy as the result of mathegrammatical illusions. The world, the universe, may be more complex than we can know or imagine, but that complexity does not include transcendental or supernatural features.
- The Subscendental Fallacy (the logician's fallacy) is that there are "deep structures" or "immanent structures" that are the locus of explanations for language, thought, and human behavior in general. Such "structures" are as ephemeral and ethereal as transcendental and supernatural worlds. They lead to conceptions of logic, mathematics, and language as "free standing," "independent," "history, culture, and value free" statements. And they support misguided sociobiological and genetic explanatory strategies.
- The Private Worlds Fallacy (the philosopher's fallacy) is that individual human beings harbor intrinsically private experiences. The profoundly social nature of humans, of symbols, and of language argues against intrinsically private experiences (as Wittgenstein, Goffman, and others have amply demonstrated).
- The Internal Life Fallacy. When we engage in discourses about surrogate counters, imitation, and artificial creatures that mimic, we need to remind ourselves that we are working in an arena of analogies and metaphors. Such efforts carry a high emotional charge because they take place at the boundaries of our skins. Analogy and generalization, if they can be shown to have constructive scientific outcomes, need not obligate us to embrace identity in, for example, building robots. Robots will not have to have "gut feelings" in the identical sense humans have gut feelings because they are organic machines. Even this "fact" needs to be

scrutinized. What we "feel" is given to us by our language, our conversations, our forms of talking. At the end of the day, feelings may not at all be straightforward matters of bio-electro-chemical processes. Electro-mechanical creatures will turn out to be just as susceptible to an internal life as humans once they have developed language, conversation, and forms of talk. This implies a social life and awareness. Roboticists may already have made some moves in this direction with the development of signal schemas and subsumption-based hormonal control (Arkin, 1998: 434ff.). The development of cyborgs and cybrids may make this issue moot.

- <u>The Psychologistic Fallacy</u> is that the human being and/or the human brain is/are free standing and independent, that they can be studied on their own terms independent of social and cultural contexts and forces. This is also known as the neuroistic error. It encompasses the idea that mind and consciousness are brain phenomena. Human beings and human brains are constitutively social. This is the most radical formulation of the response to this fallacy. A more charitable formulation would give disciplinary credibility to neuroscience and cognitive approaches to brain studies. These approaches might produce relevant results in certain contexts. Then there might be fruitful ways to pursue interdisciplinary studies linking the social sciences to the neurosciences.
- <u>The Eternal Relevance Fallacy</u> is that ancient and more recently departed philosophers should be important and even leading members of our inquiring conversations about social life. An act of intellectual courage is needed to rid us of Plato and Hegel. Once they are eliminated, an entire pantheon of outmoded and outdated thinkers, from Aristotle to Kant, will disappear from our radar. This move might also go a long way toward eliminating the worshipful attitude intellectuals often adopt to the more productive and visible members of their discourse communities. The caveat here is that some ancient and some modern thinkers (departed ones, as well as some who are still with us) who can be claimed for philosophy are still extremely valuable for us. Marx, Nietzsche, and Wittgenstein come immediately to mind.
- <u>The Corollary Intellectual's Fallacy</u> is that philosophers as philosophers (and psychologists as psychologists) have anything at all to tell us anymore about the social world. In the wake of the work of sociologists from Emile Durkheim (1995/1912) to Mary Douglas (1988), all the central problems of traditional and contemporary philosophy resolve into (not "reduce to") problems in sociology and anthropology.

NOTE

This chapter is based on my closing keynote address, Conference on Philosophy of Mathematical Practice, Free University of Brussels, Brussels, Belgium, October 24-26, 2002.

REFERENCES

Arkin, Ronald, 1998, *Behavior-Based Robotics,* Cambridge, MA, MIT Press.

Armstrong, D.M. 1968 *A Materialist Theory of the Mind,* London: Routledge & Kegan Paul.

Astington, J. 1996. "What is Theoretical about the Child's Theory of Mind?: A Vygotskian View of its Development," pp. 184-199 in P. Carruthers and P.K. Smith (eds.), *Theories of Theories of Mind,* Cambridge: Cambridge University Press.

Bakhtin, Mikhail M. 1981. *The Dialogic_Imagination: Four Essays.* Austin, TX: University of Texas Press.

Bakhtin, Mikhail M. 1986. *Speech Genres and Other Late Essays.* Austin, TX: University of Texas Press.

Bloor, D. (1976), *Knowledge and Social Imagery*, London, Routledge and Kegan Paul.

Breazeal, Cynthia, 2002. *Designing Sociable Robots.* Cambridge, MA: MIT Press.

Brothers, L. 1997. *Friday's Footprint: How Society Shapes the Human Mind,* New York, Oxford University Press.

Burton, Leone, 1999. *Learning Mathematics: From Hierarchies to Networks.* London, Routledge.

Byrne, R.W. 1991. "Brute Intellect," *The Sciences* May-June: 42-47.

Caporael, L 1990 "Foolish Liaisons: The 'New' AIISSK Programme," in manuscript, privately circulated, Department of Science and Technology Studies, RPI, Troy NY.

Caproael, L., R.M. Dawes, I.M. Orbell, and A.J.C. van de Kragt 1989. "Selfishness Examined: Cooperation in the Absence of Egoistic Incentives," *Behavioral and Brain Sciences,*" 12: 683-699, 734-739.

Cole, M. and Barbara Means 1981. *Comparative Studies of How People Think.* Cambridge,MA: Harvard University Press.

Collins R. 1989. "Toward a Neo-Meadian Sociology of Mind," *Symbolic Interaction* 12: 1-32.

Damasio, Antonio.1994. *Descartes' Error*, New York: Grosset/Putnam

Davidson, Iain and William Noble (1989), "The Archaeology of Perception," *Current Anthropology* 39, 2 (April): 125-55.

Douglas, Mary, 1988, *How Institutions Think,* Syracuse: Syracuse University Press.

Durkheim, E. (1961/1912), *The Elementary Forms of the Religious Life.* New York: Collier Books.

Durkheim, E. 1995/1912, *The Elementary Forms of Religious Life,* New York: The Free Press (in my view, now the preferred translation, by Karen Fields).

Ernest, Paul (ed.), 1994. *Mathematics, Education and Philosophy: An International Perspective.* London, The Falmer Press.

Ernest, Paul, 1999. *Social Constructionism as a Philosophy of Mathematics.* Albany NY, SUNY Press.

Fleck, L. 1979/1935, *Genesis and Development of a Scientific Fact* Chicago: University of Chicago Press.

Fodor, Jerry (1983), *The Modularity of Mind,* Cambridge, MA: MIT Press.
Franklin, Stan. 1995. *Artificial Life*, Cambridge, MA: MIT Press.
Geertz, C. 1973. *The Interpretation of Cultures.* New York: Basic Books.
Goody, J. 1977. *The Domestication of the Savage Mind.* Cambridge: Cambridge University Press.
Goffman, E. 1974. *Frame Analysis,* New York: Harper Torchbooks.
Gopkin, S. 1996. "Theories and Modules: Creation Myths, Developmental Realities, and Neurath's Boat," pp. 169-183 in P. Carruthers and P.K. Smith (eds.), *Theories of Theories of Mind.* Cambridge: Cambridge University Press.
Gordon, S. 1985. "Micro-Sociological Theories of Emotion," pp. 133-147 in H.J. Helle and S.N. Eisenstadt (eds.), *Micro-Sociological Theory: Perspectives in Sociological Theory.* Beverly Hills: Sage.
Gumplowicz, L. 1905, *Grundrisse der Soziologie,* Vienna: Manz.
Heelan, P. 1983, *Space-Perception and the Philosophy of Science,* (Berkeley: University of California Press.
Hesse, M. (1974), *The Structure of Scientific Inference,* Berkeley and Los Angeles: Macmillan, London and University of California Press,
Keesing, R.M. 1987. "Models, 'Folk' and 'Cultural': Paradigms Regained?," pp. 369-393 in D. Holland and N. Quinn (eds.), *Cultural Models in Language and Thought.* Cambridge: Cambridge University Press.
Lerman, S. (ed), 1994 *Cultural Perspectives on the Mathematics Classroom,* Dordrecht: Kluwer.
Levy-Bruhl, L. 1985/1926. *How Natives Think.* Princeton: Princeton University Press
Levi-Strauss, C. 1966. *The Savage Mind,* Chicago: University of Chicago Press.
MacKenzie, D. 1981. *Statistics in Great Britain 1865-1930,* Edinburgh: University of Edinburgh Press.
McCulloch, W.S. and W. Pitts, "A Logical Calculus of Ideas Immanent in Nervous Activity," Bulletin of Mathematical Biophysics 5, 1943.
Mills, C.W. 1963 "Language, Logic, and Culture," pp. 423-438 in C.W. Mills, *Power, Politics, and People*, I.L. Horowitz, ed. New York: Ballantine.
Nietzsche, Friedrich (1956/1887), *The Geneaology of Morals,* New York: Doubleday-Anchor.
Norman, D. and D.E. Rumelhart 1975. "Memory and Knowledge," pp. 3-32 in D. Norman and D.E. Rumelhart (eds.), *Explorations in Cognition* San Francisco: W.H. Freeman.
Olson, D. (1986), "The Cognitive Consequences of Literacy," *Canadian Psychology* 27: 109-21.
Pert, Candace. 1997. *Molecules of Emotion.* New York: Scribners.
Picard, Rosalind, 1997. *Affective Computing. Cambridge, MA: MIT Press.*
Priest, Stephen,1992 reissue ed., *Theories of Mind,* New York: Houghton Mifflin Co.
Restivo, S. 1983, *The Social Relations of Physics, Mysticism, and Mathematics* (Dordrecht: Kluwer Academic Publishers).
Restivo, S. 1992, *Mathematics in Society and History,* Dordrecht: Kluwer Academic Publishers.
Restivo, S. 2001, "Bringing Up and Booting Up: Social Theory and the Emergence of Socially Intelligent Robots," *Proceedings of the 2001 IEEE Conference on Systems, Man, and Cybernetics,* Tucson, AZ, October 7-10.
Rotman, Brian, 1993, *Ad Infinitum-The Ghost in Turing's Machine; Taking God out of Mathematics and Putting the Body Back in: An Essay in Corporeal Semiotics,* Stanford: Stanford University Press.

Sfard, A. (1994). Mathematical practices, anomalies and classroom communication problems. In P. Ernest (Ed.), *Constructing mathematical knowledge* (pp. 248-273).

Star, S.L. 1989. *Regions of the Brain: Brain Research and the Quest for Scientific Certainty.* Stanford: Stanford University Press.

Struik, D. "The Sociology of Mathematics Revisited: A Personal Note," Science and Society 50: 280-299, 1986.

Torrance, S. "Real-World Embedding and Traditional AI," preprint, Middlesex University AI Group, UK, 1994.

Valsiner, J. and R. van der Veer, 2000. *The Social Mind: Construction of the Idea*, Cambridge: Cambridge University Press,

Varela, F.J., E. Thompson, and E. Rosch, 1991, *The Embodied Mind: Cognitive Science and Human Experience*, Cambridge, MA: MIT.

Vygotsky, L.S. 1978. *Mind in Society.* Cambridge, MA: Harvard University Press.

Vygotsky, L.S. 1986. *Thought and Language*. Cambridge, MA: MIT Press.

Wertsch, J. 1991. *Voices of the Mind: A Sociocultural Approach to Mediated Action.* Cambridge, MA: Harvard University Press.

Wittgenstein, L. (1953). *Philosophical Investigations.* New York: McMillan.

Zajonc, R.B. 1980. "Feeling and Thinking: Preferences Need No Inferences," *American Psychologist* 35: 151-175.

Zajonc, R.B. 1984. "On the Primacy of Affect," *American Psychologist* 39: 151-175.

II

TAKING MATHEMATICAL PRACTICE SERIOUSLY

Chapter 5

INCOMMENSURABILITY IN MATHEMATICS

Otávio Bueno
University of South Carolina

Abstract: In this paper, as part of an argument for the existence of revolutions in mathematics, I argue that there is incommensurability in mathematics. After devising a framework sensitive to meaning change and to changes in the extension of mathematical predicates, I consider two case studies that illustrate different ways in which incommensurability emerge in mathematical practice. The most detailed case involves nonstandard analysis, and the existence of different notions of the continuum. But I also examine how incommensurability found its way into set theory. I conclude by examining some consequences that incommensurability has to mathematical practice.

Key words: Incommensurability, nonstandard analysis, set theory, mathematical practice.

1. INTRODUCTION

To make sense of mathematical practice, it is important to determine whether there are—or there aren't—revolutions in mathematics.[1] For depending on the existence of such revolutions, mathematical practice is understood in substantially different ways. For example, together with the existence of revolutionary changes in mathematics comes the existence of incommensurable mathematical theories. But is it possible that there is incommensurability in mathematics? If it is, we should be sensitive to radical meaning changes and changes in the extension of mathematical predicates, and we should identify the strategies developed in mathematical practice to deal with these changes.

In this paper, as part of an argument for the existence of revolutions in mathematics, I argue that, just as there is incommensurability in physical theories (see Kuhn [1962]), there is also incommensurability in mathematics.

[1] For several perspectives on the issue of revolutions in mathematics, see Gillies (ed.) [1992].

B. Van Kerkhove and J.P. van Bendegem (eds.), Perspectives on Mathematical Practices, 83–105.

This may seem unexpected, for reference to mathematical terms is often taken not to change in time. But I argue that this widespread assumption about mathematics is not warranted.

To make this case, I will consider a couple of case studies. The most detailed one involves nonstandard analysis, and the existence of different notions of the continuum (see Robinson [1974] and Lakatos [1978*a*]). As Lakatos and Robinson argued, in the 19th century there were two conceptions of the continuum: a Leibnizian conception, which was rich and included infinitesimals, and a Weierstrassian conception, which was simpler and excluded infinitesimals. These different conceptions had different implications about which results were true in analysis. That is, depending on the notion of the continuum that was adopted, different results were obtained. In particular, assuming the Leibnizian conception, it follows that every convergent series of continuous functions always has a continuous limit function. But this result cannot be extended to the Weierstrassian conception.

Although both Lakatos and Robinson have discussed the existence of these two conceptions of the continuum, they failed to appreciate the implication that this fact has for the incommensurability of mathematical notions. The existence of these distinct conceptions of the continuum highlights the existence of differences in the process of fixing the reference of mathematical terms and, consequently, differences in the results that are true in analysis.

I will also discuss, much more briefly, how incommensurability found its way into set theory. When Zermelo developed the first axiomatic formulation of set theory (see Zermelo [1904], [1908*a*] and [1908*b*]), he was explicitly trying to vindicate Cantor's conception of set. What he produced, however, was a significantly different notion of set—a broader and more abstract notion than Cantor's. Despite the fact that Zermelo was able to get some of the results that Cantor was trying to establish (such as the well-ordering theorem), the results were established in a framework that incorporated a notion of set that was non-Cantorian in significant ways (e.g. Cantor would never accept a unrestricted power set axiom; see, e.g., Lavine [1994]).

To make sense of the existence of incommensurable notions in mathematics, it's crucial to have an account of mathematical practice—an account that is more sensitive to meaning change as well as change in the extension of mathematical predicates. I'll sketch below a framework that should help making sense of these aspects of mathematical practice—highlighted in the case studies under discussion—and that should be general enough to be applicable to other cases. As will become clear, the framework incorporates important insights provided by Lakatos (in particular, Lakatos [1976] and [1978*c*]) as well as by Azzouni [1994]. I'll start by outlining the framework, and examining the bearing of the incommensurability issue on mathematical practice. This will pave the way to the discussion of the case studies later in the paper.

2. THE INCOMMENSURABILITY PROBLEM

What is the problem of incommensurability of mathematical theories? Briefly, the problem is that different mathematical theories—about the same domain (for example, the domain of sets)—may characterize the relevant objects in radically different ways. And there is no common standard to decide which of the two characterizations (if any) is the correct one. Depending on the framework that is assumed, a different answer regarding the correctness of the characterization is provided. If we assume the framework of one theory, we get one answer; using the framework of the other theory, we get a different one. And there's no common measure to decide between them.

Consider, for example, two different set theories: Zermelo-Fraenkel (ZF) and von Neumann-Bernays-Gödel (NBG). The former quantifies *only* over sets and is *not* finitely axiomatizable; the latter quantifies not only over sets but also over proper classes, and *is* finitely axiomatizable. Despite these differences, the two set theories are equiconsistent (that is, ZF is consistent if, and only if, NBG is also consistent). Given ZF and NBG, which sets are we referring to when we refer to *sets*? How can we distinguish ZF *sets* from NBG *sets*?[2] It's only *in the context of each set theory* that one can try to make such a distinction. But sets play exactly the same role in each theory. Given the equiconsistency between the two theories, there will always be interpretations of the formalism of the theories in which counterparts to *sets* in one theory play the role of *sets* in the other theory. 'Set' thus becomes indeterminate between ZF sets and NBG sets. There are no *common* standards to determine to which sets we are referring when we refer to *sets*. ZF and NBG are incommensurable.

But if there are incommensurable mathematical theories, what is the significance of this fact? What difference would it make to mathematical practice, and to our understanding of this practice? I think the existence of incommensurability in mathematics—similarly to its corresponding counterpart in science—has significant implications, not only to the practice of mathematics, but, in particular, to our understanding of the latter. First, if there are, for example, incommensurable theories of sets (as I think there

[2] Of course, I'm not considering *proper classes* that are not quantified over in ZF.

are), the mathematician cannot claim that the *sets* characterized in one set theory are actually *the same* as those characterized in another theory. The fact that both theories have enough in common to be considered theories of *sets* is not enough to guarantee that they characterize the *same* objects. (It's not even clear that the theories in question characterize the same *type* of objects, given that some set theories quantify over objects that are *not* sets.) Similarly, the fact that classical and relativistic mechanics have enough in common to be considered *mechanics* is not enough to guarantee that they characterize, and refer to, the same objects: Newtonian mass is not dependent on velocity, Einsteinian mass is.

Moreover, if the mathematician cannot claim that the sets studied in different set theories are different (or the same), how can he or she *identify* such sets with each other? That is, how can the mathematician identify sets across different set theories? It might be completely arbitrary to carry out such identification, and as a result, the nature of the sets in question is left unspecified. For example, it won't be determined whether an acceptable set theory should distinguish sets and proper classes. The answer would ultimately depend on the set theory one accepts, but ZF and NBG provide conflicting approaches to the issue.

Problems about identification of mathematical objects across different theories have motivated the development of certain forms of structuralism about mathematics. According to Michael Resnik [1997], for example, there is no fact of the matter as to whether objects in different mathematical structures are the same or not. The mathematical theories in question—and the resulting structures—are simply *silent* about the issue. As long as we follow the structuralist's insight that all that matters in mathematics are the *structures* in question (rather than the *objects* that exemplify such structures), the fact that we might be unable to identify the objects is no longer significant. Objects, as it were, "drop out" of the picture.

This means, in turn, that the structuralist may end up being committed to the incommensurability of mathematical notions—at least at the level of mathematical *objects*. The problem, however, is that once this first step is taken, the incommensurability will be extended to the level of *structures* as well. After all, if there is no way of identifying mathematical objects across different mathematical structures, there's no way of identifying the corresponding structures either. Any identification of structures presupposes, at least, the existence of some mapping from the *objects* of one structure to the *objects* of the other. But, given the emphasis on objects, this is exactly the presupposition that the structuralist *doesn't* grant. Moreover, even if the existence of such mappings were *stipulated*, it wouldn't guarantee that the structures in question are the same. For example, in the case of first-order structures with infinite domains, there will always be structures that satisfy exactly the same statements (they are elementarily equivalent), but which are *not* isomorphic. So there will be differences in *structure*, and the structuralist would acknowledge that. But which of these structures are we referring to? Without simply *assuming* that the structures in question can be identified, there's no way of providing a positive answer to this question. But making this assumption amounts to assuming the point in discussion.

Note that advocating the existence of incommensurability in mathematics does *not* entail that there's no genuine disagreement in mathematics. If incommensurable mathematical theories exist, then such theories may—or may not—be about *different* objects. Of course, if it turns out that the theories are *not* about the same objects, there need not be any disagreement between them. They are not rival theories after all. (Different mathematical theories may deal with different entities, or, at least, not necessarily the same ones.) But, and this is the significant case, incommensurable mathematical theories can still be about the *same* objects, in which case there might be disagreement between them. The point is only that incommensurability leaves the issue about the existence of disagreement in mathematics open. And any account of mathematical practice has to accommodate the existence of disagreement in mathematics.

But things are still more complex. After all, often mathematicians *identify* objects across different mathematical theories—e.g. by *stipulation* (see Azzouni [1994]). Making such identification often simplifies the description and the development of mathematical theories. In this way, connections across different mathematical theories are established, and information from different domains can be transferred. But stipulation wouldn't be enough to overcome the incommensurability problem. On what grounds is the stipulation made? Is it warranted? Any answer to these questions presupposes that we have already correctly identified the relevant objects, and that's exactly what is at issue.

3. A DILEMMA

Despite being fruitful, the attempt to identify mathematical objects from different theories seems to raise the following dilemma. Either the notions from different mathematical theories are identified or they are not. If they are *not* identified, we are unable to claim, say, that the old notions are being *explained* by the new ones—in particular, we cannot claim that whatever turned out to be right about the old notions can be recaptured by the new formalism. If the notions *are* identified, we have incoherence, given that the old and the new notions are *not the same.* After all, even though they may seem to lead to the same results in certain contexts, they have *different properties* (for example, ZF sets are not finitely axiomatizable, NBG sets are). Moreover, even in the cases in which the same results can be derived in each theory, the *meaning* of these results change from one theory to the other. Which theory should be used to interpret the results: the new or the old one? Given that sets do not have the same properties in each theory, the interpretation of the results changes when we move from the old to the new theory. As a result, the incommensurability issue returns. Thus, it's misleading to *identify* old and new notions. As noted above, examples of this scenario are found both in mathematics (different conceptions of set, different conceptions of the continuum), and in physics (Newtonian and Einsteinian notions of mass).

Someone may try to respond to this dilemma by examining the process through which mathematical objects are identified across different theories. Certainly, *stipulation* plays an important role here (Azzouni's emphasis on this point is exactly right). But, as Azzouni recognizes, mere stipulation cannot be the whole story. There are *pragmatic* reasons to stipulate the identification of certain mathematical objects with other such objects (say, natural numbers with certain sets). These reasons range from the development of more systematic, and in some cases, simpler mathematical theories to the formulation of new mathematical results. It's not surprising, then, that mathematical objects are routinely identified across different theories. Thus, if the identification process turns out to depend ultimately on stipulation, this emerges from good, pragmatic reasons.

But the identification of mathematical objects (via stipulation) is also guided by *structural* considerations. Isomorphism, and often partial isomorphism, play a significant role in motivating the chosen stipulation. What such isomorphisms allow is, as it were, the transferring of structure from one domain into another, the transferring of information about properties of certain objects into information about others. The existence of an isomorphism between two structures—or the embedding of one structure into the other—justifies the identification of certain mathematical objects based on the underlying structures.[3]

However, to stipulate that certain mathematical objects should be identified doesn't overcome the incommensurability problem. After all, despite the identification, there are still multiple instantiations of the resulting structures. Which of them are we referring to? It's *not* enough to claim (with the structuralist) that *it doesn't matter*, *as long as the same*

[3] Moves of this sort certainly lend plausibility to structuralism in the philosophy of mathematics. But structuralism doesn't provide the whole picture, given that, as noted above, (i) there is still underdetermination at the level of structures, and (ii) mathematical objects are still being identified with each other, despite the difficulties of making the nature of these objects precise.

properties are satisfied in each structure. After all, different structures may satisfy exactly the same mathematical properties (as non-standard models of arithmetic and analysis clearly illustrate). Thus, even if we stipulate that, say, natural numbers are such and such sets, there will be other models of set theory in which *different* sets will exemplify the *same* properties as the sets with which we have initially identified the numbers. Which sets are we referring to? Looking at their mathematical properties alone, there's no way to tell. And what else could we look at?

In trying to overcome this difficulty, one could examine the mechanisms of *referential access* to mathematical entities employed in mathematical practice. Is there anything in this practice that undermines the incommensurability of mathematical notions? I don't think so. After all, there is no way of linking, or tying in a unique way, mathematical notions and their referents, as the presence of nonstandard models and different interpretations of mathematical formalisms clearly indicate. This generates, as a result, *referential indeterminacy*. But referential indeterminacy leads to incommensurability, given that with the lack of any tight connection between mathematical notions and their referents, it's not clear that there is any common measure (or standard) to determine uniquely which object a given mathematical notion refers to.

Let me elaborate on this point. The mechanism of reference to mathematical objects articulated in mathematical practice emerges from the exploration of *mathematical systems*. A mathematical system is a collection of mathematical principles and a logic (often left implicit) to draw conclusions from such principles. The mathematical principles need not be formulated in any particular *formal* language (usually, they are expressed in mathematical English, or whatever extension of natural language that includes appropriate mathematical terms). Intuitively, the objects (whatever they may be) that satisfy such mathematical principles are the ones mathematicians refer to. Typically, there will be many different objects satisfying the same description. For example, in the case of (first-order) mathematical theories with infinite domains, there will always be nonstandard models of the theories in which the same results will come out true, even though the extension of the mathematical predicates of these theories is radically different from model to model.[4] Clearly, there is no other way to refer to mathematical objects but through the language of the

4 Even in the case of second-order mathematical theories, there will be such nonstandard models. As is well known, besides standard semantics, second-order logic also has Henkin models, that is, models in which the second-order variables do not range over the full power set of the universe of individuals. With Henkin models, second-order logic has the same metalogical features as first-order logic, including the existence of nonstandard models via the Löwenheim-Skolem theorem (for details, see, e.g., Shapiro [1991]).

relevant mathematical system, and it's through this language—and only through it—that we can "grasp" such objects. (Of course, there's no causal contact or instrumental access to mathematical objects.[5]) As a result, and for the reasons discussed above, there is no unique way of specifying the objects we are referring to—and incommensurability follows.

Of course, "fixing" the reference *up to isomorphism* doesn't help here, given that despite the existence of an isomorphism between two structures, the objects that are referred to in each of them can be radically different. To which of these objects are we referring? Going structuralist doesn't help either, because, as we saw, there is incommensurability at the level of structures too (e.g. different types of set instantiate the same number-theoretic properties). Hence, the incommensurability problem still stands.

Perhaps a different perspective on incommensurability in mathematics is provided by a particularly strong form of platonism: full-blooded platonism (FBP; see Balaguer [1998][6]). According to FBP, every logically possible mathematical theory is actually true of some part of the mathematical realm. The mathematical realm is a realm of plenitude, where, for example, Cantorian and non-Cantorian sets, Hilbert spaces, nonstandard models, differential equations and unusual topologies are all found "side by side", as it were. The only requirement for the existence of a mathematical entity is that it is referred to by a *consistent* mathematical theory.

Can FBP solve the problem of the incommensurability of mathematical theories? If every logically possible mathematical theory is actually true of some part of the mathematical realm, then the fact that, say, ZF and NBG provide different notions of set is no longer a problem. Each set theory will correctly describe the relevant part of the mathematical realm. The FBP-ist need not select a unified picture of the mathematical world. No matter how

[5] To use Azzouni's terminology, there's no *thick epistemic access* to mathematical entities (see Azzouni [1994] and [1997]). An epistemic access to an object is *thick* if (i) it is *robust* (e.g. you blink, you move away and the object is still there); (ii) the access can be *refined* (e.g. you can get closer for a better look); (iii) the access enables us to *track* the object in question, and (iv) we can use properties of the object in question to *get to know* other properties of the object. Mathematical objects clearly fail to satisfy (at least) conditions (i) and (iii). To the extent that we have access to such objects at all, it is only a *thin* form of access. That is, the access is *through a theory* that has the usual methodological virtues: it is simple, familiar, has large scope, is fecund, and is successful under testing. Using van Fraassen's [1980] distinction between acceptance and belief, we can say that having a thin form of epistemic access to mathematical entities gives us reason to *accept* such entities, although not a reason to *believe* in their existence (reasons for acceptance are *pragmatic* reasons, after all).

[6] It should be noted that Mark Balaguer himself is *not* a FBP-ist. He thinks that there's no fact of the matter as to whether platonism or nominalism is true. But as part of his case for the latter claim, he provides the most interesting and thorough defense of FBP.

dissimilar the mathematical objects are from one another, they will all be there, and the consistency of the mathematical theories is enough to guarantee reference to the relevant objects.

This is a very interesting proposal. But it's not clear that the proposal *solves* the incommensurability problem. Rather, it *embraces* the issue raised by the incommensurability of mathematical notions. For the FBP-ist, *different* (consistent) mathematical *theories* are about *different* mathematical *objects*. But to embrace this conclusion, the BFP-ist seems to be committed to the *impossibility* of genuine disagreement in mathematics. After all, distinct (but consistent) mathematical theories deal with distinct parts of the mathematical realm. So, the theories are about different objects, and hence, strictly speaking, they cannot be in conflict with each other. The problem, however, is that, as a feature of mathematical practice, *there seems to be* genuine disagreement in mathematics. Even a cursory look at the history of mathematics makes this clear. Mathematicians systematically disagree about a number of issues, from the introduction of new mathematical entities (consider the debates surrounding the formulation of irrational numbers, imaginary numbers, or different notions of set), through the appropriate standards of proofs (are informal arguments enough?), to the adequacy of mathematical techniques (consider the debate between constructivists and classical mathematicians). Without being able to make sense of these debates *at least for what they seem to be*—that is, disagreements about how to conceptualize the objects and standards in question—it is not clear that FBP provides an adequate picture of mathematical practice.[7] As a result, in the end, FBP doesn't yield an acceptable account of the incommensurability of mathematical notions.

4. A SIMPLE PATTERN

To examine the case study about incommensurability in mathematics, it will be useful to have a pattern to guide the discussion. To do that, I'll use some ideas that Lakatos put forward for a different purpose.

[7] Of course, the FBP-ist can always claim that the disagreement among mathematicians is simply an *appearance*; in fact, there is no such thing as a controversy in mathematics. Mathematicians that seem to disagree are simply describing different aspects of the mathematical universe. Even if this were the case, the FBP-ist owes us at least an explanation as to *why* mathematicians *seem to be* disagreeing if they are really talking past each other. But it's not clear that the FBP-ist has an adequate explanation—sensitive to mathematical practice—to offer here. Inflating the ontology, by requiring the existence of all mathematical objects that logically possibly could exist, does all the work for the FBP-ist. This move is simply silent with respect to mathematical practice.

In a very interesting piece on mathematical reasoning, Lakatos discusses three types of mathematical proofs: *pre-formal proofs*, *formal proofs* and *post-formal proofs* (Lakatos [1978*b*]). Most of mathematical practice is carried out in terms of pre- or post-formal proofs, given that the bulk of mathematical reasoning is done in a somewhat informal way. In particular, the underlying logic typically is hardly ever made explicit, and hence inferences are drawn informally. Unless the subject under consideration is particularly controversial, or there are genuine doubts about the consistency of the theories in question, the proofs provided are never formal.

Lakatos [1978*a*] also develops a simple (Hegelian) pattern of development of mathematical conjectures, and this pattern helps to illuminate important features of mathematical practice. First, an initial conjecture is advanced (*thesis*). Second, attempts are made both to *prove* the conjecture and to *refute* it (*antithesis*). Finally, as the result of this double-headed process, a refined conjecture is obtained and eventually established. This new conjecture incorporates what was learned from possible *counterexamples* to the original conjecture as well as the *new concepts* introduced in the attempt to formulate and establish the refined conjecture (*synthesis*).

This simple pattern beautifully illustrates the development of several mathematical results. For example, the formulation and eventual proof of the well-ordering theorem in set theory can be clearly described in terms of the pattern. Let me briefly sketch why this is the case.

As part of his study of trigonometric series, Cantor was led to provide a formulation of real numbers in terms of sequences of rational numbers, and to study infinite iterations of certain operations over collections of real numbers. This latter study led him to develop a mathematical theory of the infinite. To develop that theory, Cantor needed a mathematical principle to justify carrying over properties of finite numbers into the transfinite. This idea—of transferring some properties from finite domains to infinite ones—provided a guiding heuristic motivation for the construction of Cantor's theory (see Hallett [1984]). The mathematical principle Cantor introduced, which clearly holds for finite numbers, was the well-ordering principle. According to this principle, "it is always possible to bring every *well-defined* set into the form of a *well-ordered* set" (Cantor [1883], p. 550; see Kanamori [1996], p. 5).[8] Interestingly enough, Cantor considered this principle "fundamental", since it is "rich in consequences and particularly

[8] According to Cantor [1883], a set M is *well-ordered* by a relation $<$ if: (a) M is linearly ordered by $<$; (b) M has a first element m_0 (according to the relation $<$); and (c) whenever $N \subseteq M$ and $\exists m \in M\text{-}N$ such that $\forall n \in N\ (n < m)$, then there is a $<$-smallest $m \in M\text{-}N$ such that $\forall n \in N\ (n < m)$. This definition is equivalent to the standard definition. (For a discussion, see Hallett [1984], pp. 51-52.)

remarkable for its general validity". On his view, the principle was actually a "law of thought". It clearly yielded the justification for unifying the finite and the infinite that Cantor needed. But, alas, Cantor never gave a *proof* of the principle (why should one try to prove a "law of thought" anyway?).

In 1904, at the Third International Congress of Mathematicians in Heidelberg, König presented what he took to be a *counterexample* to Cantor's principle, offering an argument to the effect that there is *no* well-ordering of the continuum. If this result were correct, it would of course have seriously damaged Cantor's program. But it turned out that König's "proof" contained a mistake, and he had to withdraw it. The error was found by Zermelo, who showed that König relied on a theorem of Bernstein that was incomplete in the relevant case.[9]

Motivated in part by this incident, and in order to provide support for Cantor's case, Zermelo eventually proved a theorem to the effect that *every set* can be well-ordered, the well-ordering theorem (Zermelo [1904]; for a discussion, see Kanamori [1996], pp. 10-13, and Kanamori [1997]). But in order to prove this theorem, Zermelo had to use what was soon to be called the Axiom of Choice (AC).[10] It is only based on this axiom that the theorem holds. On Zermelo's view, however, AC was nothing but a "logical principle", and although it cannot "be reduced to a still simpler one", it is "applied without hesitation everywhere in mathematical deduction" (Zermelo [1904], p. 141). A crucial point here is that AC is consistent with Cantor's requirement that the finite and the transfinite should be approached uniformly, since the axiom extends to infinite sets an unproblematic feature of finite sets (Kanamori [1996], p. 10; see also Moore [1982], and Lavine [1994], pp. 103-118). However, in introducing AC, and in developing his approach to set theory, Zermelo ended up formulating not exactly a Cantorian view about sets. Zermelo's notion of set is much more abstract and general than Cantor's, given that it includes, in particular, an unrestricted power set axiom, something Cantor would never endorse (see, e.g., Lavine [1994]).[11]

This case can be clearly described in terms of Lakatos' pattern. Cantor's initial conjecture (the well-ordering principle) is the *thesis*, the initial conjecture that Cantor never actually proved. The *antithesis* is given by

9 For a discussion of this episode, see van Heijenoort (ed.) [1967], p. 139; Moore [1982], pp. 86-88; Lavine [1994], p. 103; and Kanamori [1996], pp. 9 and 49, note 22.

10 In one of its formulations, this axiom states that: If T is a set whose elements are sets that are different from $\varnothing$ and mutually disjoint, its union $\cup T$ includes at least one subset S having one and only one element in common with each element of T (see Zermelo [1908*b*], p. 204, and Moore [1982], pp. 5-11, and 321-334).

11 I discuss this episode further in Bueno [2000] and Bueno [2002], but I draw a different moral in the present context.

König's attempt to refute Cantor's conjecture, trying to establish that the continuum cannot be well ordered. The *synthesis* is finally obtained with Zermelo's proof of the well-ordering theorem. But given the use of AC in Zermelo's proof, and the abstract treatment of sets articulated in his axioms for set theory, the meaning of Cantor's original conjecture is changed. After all, Cantor's original "law of thought" is only valid assuming a particularly strong principle about sets (namely, AC). In other words, as opposed to Cantor's initial conjecture, it's *not* the case that every set can be well ordered *in general*. Every set can be well ordered—*as long as AC holds for such sets*. Now, whether the latter condition is, or is not, satisfied depends, ultimately, on the context (the mathematical system) one considers, and the type of sets in question. (As is well known, the properties of sets differ quite dramatically in the presence of AC.) In the attempt of refining and proving Cantor's initial conjecture, Zermelo ended up establishing a *different*, *more restricted* result.

But Lakatos' simple dialectical pattern needs to be used with care. With the possibility of incommensurable mathematical theories, the result established in the final synthesis can be more than a simple *refinement* of the original conjecture. As the discussion above of Cantor's case illustrates, the final result may mean something *different* from what was meant with the initial conjecture. After all, with Zermelo's work, a more abstract notion of set is in place. But note that this is not a problem. It's actually an *advantage* of this pattern, given that it is able to accommodate the development of richer and more complex bodies of mathematics. What emerge from the interaction between the initial conjecture and its refinement are new concepts and mathematical results that, in the end, are more interesting.

There's no doubt that the simple pattern described by Lakatos accommodates an important feature of mathematical practice: the interplay between attempts to prove and refute a given conjecture, leading to the formulation of new notions and the refinement of the initial guess. But given that the original conjecture and the refined one may not have the same meaning, the original conjecture *might be right in the context in which it was originally formulated*, and the final theorem may actually establish a *different result* than the one that was initially formulated. This seems to introduce a lack of continuity in the development of mathematical theories.

Thus, and Lakatos certainly emphasizes this point, we have a challenge to the received view about the history of mathematics, according to which the development of mathematics is the result of the accumulation of truths about mathematical objects. Rather, a closer look at the history of mathematics reveals that radical discontinuities are indeed possible. To accommodate this possibility, Lakatos' pattern does *not* require a tight connection between the initial conjecture and the final theorem. Once the process of concept revision and refinement is in place, there is always the possibility of radical meaning change.

5. INCOMMENSURABILITY IN NONSTANDARD ANALYSIS

The impact of the incommensurability of mathematical notions in mathematical practice is not restricted to set theory. To consider an additional illustration of the incommensurability phenomenon, I'll examine the construction of nonstandard analysis by Abraham Robinson, who, among other achievements, put infinitesimals back into calculus (see Robinson [1974], [1979*a*] and [1979*b*]). The reason for focusing on nonstandard analysis is that it provides an interesting example of the bearing of the incommensurability issue on mathematical practice—at least in the case of Robinson's work. In this way, we can appreciate the overall relevance of the issue.

It is well known that the early formulation of the calculus, due to Leibniz and Newton, was heavily dependent on infinitesimals, which were employed, for instance, in the derivation of the rules of Newton's method of fluxions (see Lavine [1994], pp. 15-26). Intuitively, infinitesimals were taken to be indefinitely small quantities, smaller than any finite quantity. But lacking a precise mathematical definition, infinitesimals were received with harsh criticism. Discussing Newton's method of fluxions, Berkeley pointed out:

> And what are these fluxions? The velocities of evanescent increments? And what are these evanescent increments? They are neither finite quantities, nor quantities infinitely small, nor yet nothing. May we not call them the ghosts of departed quantities? (Berkeley [1734], p. 88)

As to Leibniz's infinitesimals, Berkeley was no less caustic:

> [Our modern analysts] consider quantities infinitely less than the least discernible quantity; and others infinitely less than those infinitely small ones; and still others infinitely less than the preceding infinitesimals, and so on without end or limit. (Berkeley [1734], p. 68)

Now, Leibniz was trying to devise a program of construction of numbers that would include infinitesimals in a suitable way (see Leibniz [1701]). The idea was to introduce the latter, by appropriate arithmetic rules, as ideal numbers into the system of real numbers, in such a way that the resulting system would have the same properties as the real number system. Nevertheless, given that neither Leibniz nor his followers managed to produce such a system, infinitesimals gradually fell into disrepute, and were eventually eliminated in the classical theory of limits elaborated in the nineteenth century (see Lavine [1994], pp. 26-41; see also Robinson [1974], pp. 260-282).

However, as Lakatos would say, with sufficient heuristic resources every program can be revived. And in 1960, with the work of Robinson, Leibniz's program was brought back. In fact, what Robinson argued ([1974], p. xiii) is that the model-theoretic techniques developed in the 20th century provided the adequate framework in which Leibniz's intuitions could be properly articulated and vindicated. Robinson showed that the ordered fields that are nonstandard models of the theory of real numbers could be thought of, in the metamathematical sense, as non-archimedean ordered field extensions of the reals,[12] and they included numbers *behaving like infinitesimals* with regard to the reals. Moreover, since these ordered fields are *models* of the reals, they have the same properties as the latter. As a result, Leibniz's problem was solved.

The crucial notion Robinson used to provide nonstandard models of analysis was that of *enlargement*.[13] Given a structure **R** (say, the real number structure), an enlargement ***R** of **R** is an expansion of **R** (in technical parlance, **R** is a substructure of ***R**), such that a sentence α is true in **R** if and only if α is also true in ***R** (that is, **R** and ***R** are elementarily equivalent). In other words, an enlargement B of a given structure A is an extension of A that preserves the truth-values of the sentences that hold in A. The decisive result from model theory that Robinson employed in the development of nonstandard analysis was the compactness theorem, according to which if K is a set of sentences such that every finite subset of K is consistent, then K is also consistent. What the compactness theorem allowed him to prove is that for any structure A there is an enlargement B of A.[14] However as Robinson

12 A field over the real numbers is *archimedean* if for any pair of real numbers a and b, $0 < a < b$, there exists a natural number n (in the *ordinary*, *standard* sense) such that $b < na$. This postulate is not true in Robinson's nonstandard models, and in this sense, the latter are non-archimedean (see Robinson [1974], pp. 266-267).

13 Robinson's original formulation of nonstandard analysis, in 1960, was articulated in type theory (for an overview, see Robinson [1974]). His account was subsequently reformulated by Luxemburg [1962], using an ultrapower of the first-order structure of real numbers, and by Machover [1967] in set-theoretic terms (the latter account is also presented in Bell and Machover [1977], pp. 531-575, which provides a concise and helpful introduction to nonstandard analysis). Certainly the last two formulations, especially Luxemburg's, were crucial for spreading nonstandard analysis among mathematicians, who typically are not used to the background in logic required by Robinson's type-theoretic version (see Dauben [1995], pp. 393-396).

14 Given Robinson's use of type theory, a point should be noted. It is well known that type theory is only complete and, in particular, compact if we consider Henkin models. Again,

pointed out, the enlargement is by no means unique. The real number structure **R** has several enlargements ***R**, and any of them provides a nonstandard model of analysis. However, once an enlargement has been chosen, "the totality of its internal entities is given with it" (Robinson [1967], p. 29). As a result:

> corresponding to the set of natural numbers **N** in **R**, there is an internal set ***N** in ***R** such that ***N** is a proper extension of **N**. And ***N** has "the same" properties as **N**, i.e. it satisfies the same sentences of *L* just as ***R** has "the same" properties as **R**. ***N** is said to be a *Non-standard model of Arithmetic* [just as ***R** is called a *nonstandard model of Analysis*]. From now on all elements (individuals) of ***R** will be regarded as "real numbers", while the particular elements of **R** will be said to be *standard*. (Robinson [1967], pp. 29-30).[15]

Now, the crucial feature of ***R** is that it is a *non-archimedean* ordered field. Therefore, ***R** contains infinitely small numbers (infinitesimals), that is, numbers $a \neq 0$ such that $|a| < r$ for all standard positive r (Robinson [1967], p. 30).

Since the structures **R** and ***R** satisfy the same set of sentences, the properties of relations and functions in one structure can be "transferred" back into the other, and vice-versa. This provides the main heuristic move used by Robinson, namely *transfer principles*. These principles are straightforward consequences of the model-theoretic framework in which Robinson worked, given the elementary equivalence of the structures under consideration. In other words, there are significant (model-theoretic) interconnections between **R** and ***R**, and the decisive trait of nonstandard analysis is to explore them. Although we may not know whether a given result holds in **R**, by embedding it into ***R**, we have "more structure" to

in these models, the interpretation of the quantifiers takes into account not the totality of all relations of a given type, but only a subset of them. These are the so-called *internal relations*. As Robinson points out, "if *S* is a set or relation in **R** then there is a corresponding internal set or relation **S* in ***R**, where *S* and **S* are denoted by the same symbol in *L* [the higher-order language in which statements about the structure **R** are formulated]" (Robinson [1967], p. 29). However, Robinson insists, "not all internal entities of ***R** are of this kind" (*ibid.*). Indeed, since there is an infinite set in **R**, there is a set in ***R** which contains an internal relation which is not a standard relation, that is, which is not denoted by any constant in the fixed set of sentences *K* (see Robinson [1974], pp. 42-45).

15 Note Robinson's use of quotation marks when he describes the relation between the nonstandard models ***N** and ***R** and their standard counterparts **N** and **R**. The quotations highlight the fact that ***N** and **N** (as well as **R** and ***R**) are actually *not* the same—as they couldn't be, given that they have different elements—even though they satisfy the same sentences. Different extensions for the same predicates are found in each model.

work with, and in this way, we may be able to establish the result. Using a transfer principle, we then establish that this result also holds in **R**.

By systematically exploring this heuristic strategy, Robinson was not only able to simplify several proofs of established theorems, but also to prove new results. For example, in a joint work with Bernstein, he solved an invariant subspace problem, using nonstandard techniques. This was an open problem, which hasn't been solved yet with the resources of classical analysis (see Bernstein and Robinson [1966]).[16] Moreover, he also provided results in general topology, theory of distributions, topological and metrical groups, applied mathematics etc. (see Robinson [1974]). Robinson also showed how analysis could be reformulated with infinitesimals. For example, he established that the real-valued function f is continuous at x_0 in the real number structure **R** if and only if $f(x_0 + \eta)$ is infinitely close to $f(x_0)$ in the nonstandard model ***R**, that is, $f(x_0 + \eta) - f(x_0)$ is infinitesimal, for all infinitesimal η (see Robinson [1967], pp. 30-31, and Robinson [1974], pp. 49-88). In other words, a completely new approach was devised.

Robinson's formulation of nonstandard analysis was not only heuristically fruitful (given, for example, the solution of the invariant subspace problem), but it also led to a re-evaluation of the history of mathematics, in particular of the calculus (see Robinson [1974], pp. 260-282, and Robinson [1967]). After all, until the construction of nonstandard analysis, the history of mathematics has been written on the assumption that infinitesimals were, in Cantor's words, "the *cholera bacillus* of mathematics",[17] or in Berkeley's view, "the ghosts of departed quantities" (Berkeley [1734], p. 88). As a result, the insightful ideas of Leibniz and many others about infinitesimals were disregarded as mere eccentricities. As Robinson points out, with nonstandard analysis it's possible to devise a more sympathetic evaluation.

[16] Bernstein and Robinson proved the following theorem: if T is a bounded linear operator on an infinite-dimensional Hilbert space H over the complex numbers, and if $p(z) \neq 0$ is a polynomial with complex coefficients such that $p(T)$ is compact, then T leaves invariant at least one closed subspace of H other than H or $\{0\}$ (Bernstein and Robinson [1966], p. 421). The main idea is to associate with the Hilbert space H a larger space $^*H$ which, given the construction of the enlargement, has the same properties as H. The problem is then solved by considering the invariant subspaces in a subspace of $^*H$, whose number of dimensions is a nonstandard positive integer (*ibid.*). After seeing the nonstandard proof of this result, Halmos obtained a proof using standard techniques, essentially translating Bernstein and Robinson's model-theoretic argument into ordinary mathematics (see Halmos [1966], and for a discussion, Halmos [1985], pp. 204 and 320, and Dauben [1995], pp. 327-329).

[17] Cantor expressed this evaluation in a letter to the Italian mathematician Vivanti, December 13, 1893 (see Dauben [1979], p. 233).

As an illustration of the historical fertility of nonstandard analysis, Robinson considered the celebrated case of Cauchy on (uniform) continuity (see also Lakatos [1978*a*]). In 1821, Cauchy proved a theorem to the effect that the sum of a convergent series of continuous functions was continuous. In the proof, although not in the statement of the theorem, Cauchy made full use of infinitesimals. Five years later, Abel provided an example of a convergent series of continuous functions whose sum was *not* continuous, namely $\sin x + 1/2 \sin 2x + 1/3 \sin 3x + \ldots$ It turns out that Cauchy's theorem would be entirely *correct* if we consider that the convergence holds in an interval containing *not only standard, but also nonstandard points* (that is, points of the form $x+\eta$, where x is real and η infinitesimal). In other words, Cauchy held a richer conception of the continuum than Abel (and later Weierstrass), since it included infinitesimals. In fact, Robinson showed that if nonstandard points are introduced, uniform convergence of an infinite series in an interval—that is sufficient for the continuity of the sum—is equivalent to pointwise convergence. In this way, in the light of nonstandard analysis, we can understand why Cauchy presented his theorem, and why it is inadequate to read it assuming the Weierstrass continuum that excludes infinitesimals.[18]

In other words, in the 19th century, there were two conceptions of the continuum, and they yielded very different results regarding the properties of real numbers. The conceptions were actually incommensurable in that (i) depending on the conception one adopts, the *meaning of 'real number' is different* (e.g. one conception includes infinitesimals, the other doesn't), and (ii) there is *no common standard* to assess the adequacy of the different notions (after all, different results about the continuity of series of functions are established by each conception).

The existence of these different conceptions changes the practice of mathematicians. The Leibnizian mathematician (such as Cauchy) has more structure to work with, given that the Leibnizian continuum incorporates infinitesimals and infinitely large numbers. As a result, he or she is able to obtain results that couldn't be obtained on the Weierstrassian conception of the continuum. The latter conception, despite being less rich, had the initial benefit of not depending on the introduction of infinitesimals. But this came with a cost: some results about real numbers *cannot* be obtained in this framework, such as Cauchy's result regarding convergent series.

Now, it's important to note that Robinson's notion of the continuum, although being in spirit close to Leibniz's, is in important respects *different* from Leibniz's too. The fact that there are objects that 'behave like

[18] For details, see Robinson [1967], pp. 35-38, Robinson [1974], pp. 269-276, Lakatos [1976], pp. 127-141, Lakatos [1978*a*], and Dauben [1995], pp. 349-355.

infinitesimals', or 'could be taken to be infinitesimals' in an enlargement *doesn't entail* that these objects *are* infinitesimals. As Robinson acknowledges, nonstandard analysis provides a framework to *represent* infinitesimals. This is *very different* from actually incorporating or invoking infinitesimals *as Leibniz conceived of them*. The model-theoretic framework that Robinson employed lets reference to mathematical objects completely loose. The *plasticity* provided by the various reinterpretations of mathematical principles that the model-theoretic approach offers is both the strength and the weakness of the program. It's the strength in that a rich variety of structures is generated in this way, and several results are then obtained. It's the weakness in that, on the model-theoretic account, there's no close tie between mathematical notions and their referents. It's always possible to find deviant interpretations of the formalism of analysis in which the relevant results come out true, even though the extension of the predicates in question changes dramatically. This means that there's no way in which Robinson could claim that the objects that play the role of infinitesimals in his framework are *Leibnizian infinitesimals*, even though such objects may share some of the properties of the latter. There is incommensurability even here.

But what exactly are Leibnizian infinitesimals? To the extent that Leibniz himself was clear about the nature of these entities, the constraints he imposed on them were minimal. Clearly, he assigned an instrumental role to infinitesimals, and was explicit in taking them only to be "useful fictions" (Leibniz [1716]). That doesn't entail, of course, that infinitesimals could be anything one pleases. They had some properties, lacked others, and had otherwise to behave in certain ways. But whatever Leibniz and his followers had in mind when they entertained the notion of infinitesimal, it was certainly *not* a particular kind of object in a Henkin model that is an enlargement of certain type-theoretic structures. This is further grist for the incommensurability theorist's mill.

The incommensurability, though, is not restricted to Leibniz's and Robinson's view of the continuum. It also applies to Robinson's framework and standard analysis. It was initially thought—and Robinson certainly presented the issue in this way—that nonstandard analysis is simply a conservative extension of standard analysis, and so any result obtained via nonstandard techniques can also be obtained via the standard approach. In this sense, although useful, nonstandard analysis is ultimately *dispensable*. It turns out, however, that things are not so simple. As Henson and Keisler [1986] have shown, there are theorems that can be proved with nonstandard analysis but which cannot be proved without it. The source of the confusion about the issue is the following. On the one hand, we have the correct statement (essentially the one made by Robinson): if a theorem can be proved using nonstandard analysis, it can be proved in Zermelo-Fraenkel set theory with the Axiom of Choice, and therefore it is acceptable as a theorem of (standard) mathematics. However, from this it *doesn't* follow that there is no need for nonstandard analysis. As Henson and Keisler indicate, almost all results in classical mathematics employ methods available in second-order arithmetic, given appropriate comprehension and choice axioms. In other words, mathematical practice is articulated in a conservative extension of second-order arithmetic, and the higher level of sets available in ZFC is not used. What the authors then show is that nonstandard analysis (that is, second-order nonstandard arithmetic) with a saturation principle (often used in nonstandard arguments) has the same strength as *third-order* arithmetic. Therefore, "there are theorems which can be proved with nonstandard analysis but cannot be proved by standard methods" (Henson and Keisler [1986], p. 377).

This means that the frameworks of standard and nonstandard analysis are importantly *different*, particularly in the context of everyday mathematical practice. How could we choose between them? Again, there's *no common standard* to make the choice. One could prefer the standard approach because it's familiar and does not invoke "suspicious" entities like infinitesimals. But these considerations clearly *presuppose* the *adequacy* of the standard account and the *inadequacy* of the way infinitesimals are characterized in nonstandard analysis. As a result, the considerations simply beg the question. One could prefer, instead, the nonstandard approach because it's richer and can be used to obtain new results. But this also begs the question, now against the standard approach. The alleged new results from nonstandard analysis might be obtained using standard techniques, in which case the results are not exactly *new* and the nonstandard framework is not necessarily *richer* than the standard one.[19]

This is *not* an argument against nonstandard analysis, of course. It's only an argument for the incommensurability of the standard and the nonstandard approaches. This fact has a significant impact on mathematical practice. Given the strength of nonstandard techniques (and the strength of the mathematical framework they presuppose), the incorporation of nonstandard

[19] How can one claim that the alleged new results from nonstandard analysis are actually results in *standard* analysis? An argument (which I don't exactly endorse) could go like this. If the new results are established using resources *beyond* the scope of the standard framework, they simply do *not* belong to the latter. If the new results are established based on standard resources, they can be obtained *without* nonstandard techniques. So, in either case, there's no need for the nonstandard approach. What is established properly is established using the standard approach. (Of course, this argument fails to note that the results in question may have different meanings in the standard and the nonstandard frameworks.)

techniques into everyday mathematical practice would increase the overall strength of the mathematical systems used in that practice. The rejection of these techniques, in turn, would amount to an important loss to mathematical practice. In either case, important changes emerge.

Finally, note that the development of nonstandard analysis illustrates the simple dialectical pattern discussed by Lakatos. Leibniz's initial conjecture about the possibility of developing a system of real numbers incorporating infinitesimals was the *thesis*. The difficulties of actually implementing such a system, which eventually led to the exclusion of infinitesimals from the calculus with Weierstrass, illustrate the *antithesis*. Robinson's work on nonstandard analysis, that motivated a reassessment of the whole debate, eventually provides the *synthesis*. Interestingly enough, the synthesis changes substantially the meaning of the original Leibnizian project, given the nature of the infinitesimals put forward by Robinson and the model-theoretic techniques he devised to achieve that. Moreover, note that throughout this development, the proofs provided have always been rather informal (in most cases, they were actually pre-formal proofs in Lakatos' sense). The arguments given have typically been rather intuitive, even in Robinson's case, despite his use of type theory. Given the way in which mathematical practice is actually conducted, that's the expected outcome.

6. CONCLUSION

If the considerations above are correct, a case for the incommensurability of mathematical notions can be made. Incommensurability is not restricted to scientific theories after all, but is also part of mathematical practice. Theory change in mathematics, just as theory change in science, becomes a more complex, more interesting and not a cumulative phenomenon. As with science, in mathematics sensitivity to meaning change is required. This means that a simple cumulative pattern of mathematical development doesn't seem to make sense of mathematics. This, in turn, opens up the possibility for the existence of revolutions in mathematics—radical cases of mathematical theory and conceptual change. In other words, as Lakatos would put it, the usual Euclidean model of mathematical development, according to which mathematical theories are a body of eternal, immutable truths, has to go. A more fine-grained model is in order.

While I have no intention of providing such a model here—the aim of the paper is only to make the initial case for the incommensurability of certain mathematical notions—I'd like to conclude by highlighting the impact that incommensurability has on mathematical practice. Mathematicians are typically sensitive to meaning changes, and as part of their practice, they develop strategies to track such changes. One simple strategy is to identify whether new results can be established once a new conceptual setting is advanced. If new results can be obtained, this provides *prima facie*, but certainly not conclusive, evidence for meaning change. But, alas, as the considerations above indicate, things are never that simple. There may be meaning change even when *the same results* are established in the old and the new frameworks. The case of nonstandard analysis beautifully illustrates this. On the usual reading (the one favored by Robinson), given that nonstandard analysis is essentially a conservative extension of standard analysis, all new results proved by nonstandard techniques could always be obtained without the latter, but they need not have the same meaning (given the quantification over infinitesimals in one framework and their absence in the other). So meaning change need not be correlated with the presence, or not, of new results.

This leaves the incommensurability of mathematical notions as a feature—an important feature—of mathematical practice. The existence of incommensurability, in turn, makes mathematical practice a much more interesting, rich and complex phenomenon to study. In the end, there's much more to that practice than meets the eye.

REFERENCES

Azzouni, J. [1994]: *Metaphysical Myths, Mathematical Practice*. Cambridge: Cambridge University Press.

Azzouni, J. [1997]: "Thick Epistemic Access: Distinguishing the Mathematical from the Empirical", *Journal of Philosophy 94*, pp. 472-484.

Balaguer, M. [1998]: *Platonism and Anti-Platonism in Mathematics*. New York: Oxford University Press.

Bell, J.L., and Machover, M. [1977]: *A Course in Mathematical Logic*. Amsterdam: North-Holland.

Berkeley, G. [1734]: "The Analyst", in Berkeley [1951], vol. 4., pp. 65-102.

Berkeley, G. [1951]: *The Works of George Berkeley Bishop of Cloyne*. (Edited by A.A. Luce and T.E. Jessop.) London: Thomas Nelson and Sons.

Bernstein, A., and Robinson, A. [1966]: "Solution of an Invariant Subspace Problem of K.T. Smith and P.R. Halmos", *Pacific Journal of Mathematics 16*, pp. 421-431. (Reprinted in Robinson [1979*b*], pp. 88-98.)

Bueno, O. [2000]: "Empiricism, Mathematical Change and Scientific Change", *Studies in History and Philosophy of Science 31*, pp. 269-296.

Bueno, O. [2002]: "Mathematical Change and Inconsistency: A Partial Structures Approach", in Meheus (ed.) [2002], pp. 59-79.

Cantor, G. [1883]: "Über unendliche, lineare punktmannigfaltigkeiten. V", *Mathematische Annalen 21*, pp. 545-591.

Dauben, J.W. [1995]: *Abraham Robinson: The Creation of Nonstandard Analysis, a Personal and Mathematical Odyssey*. Princeton, NJ: Princeton University Press.

Gillies, D. (ed.) [1992]: *Revolutions in Mathematics*. Oxford: Clarendon Press.

Hallett, M. [1984]: *Cantorian Set Theory and Limitation of Size*. Oxford: Clarendon Press.

Halmos, P. [1966]: "Invariant Subspaces of Polynomially Compact Operators", *Pacific Journal of Mathematics 16*, pp. 433-437.

Halmos, P. [1985]: *I Want to Be a Mathematician. An Automathography*. New York: Springer-Verlag.

Henson, C.W., and Keisler, H.J. [1986]: "On the Strength of Nonstandard Analysis", *Journal of Symbolic Logic 51*, pp. 377-386.

Kanamori, A. [1996]: "The Mathematical Development of Set Theory From Cantor to Cohen", *Bulletin of Symbolic Logic 2*, pp. 1-71.

Kanamori, A. [1997]: "The Mathematical Import of Zermelo's Well-Ordering Theorem", *Bulletin of Symbolic Logic 3*, pp. 281-311.

Kuhn, T. [1962]: *The Structure of Scientific Revolutions*. Chicago: University of Chicago Press. (A second edition was published in 1970.)

Lakatos, I. (ed.) [1967]: *Problems in the Philosophy of Mathematics*. Amsterdam: North-Holland.

Lakatos, I. [1976]: *Proofs and Refutations*. Cambridge: Cambridge University Press.

Lakatos, I. [1978*a*]: "Cauchy and the Continuum: The Significance of Non-Standard Analysis for the History and Philosophy of Mathematics", in Lakatos [1978*c*], pp. 43-60.

Lakatos, I. [1978*b*]: "What Does a Mathematical Proof Prove?", in Lakatos [1978*c*], pp. 61-69.

Lakatos, I. [1978*c*]: *Mathematics, Science and Epistemology*. Cambridge: Cambridge University Press.

Lavine, S. [1994]: *Understanding the Infinite*. Cambridge, Mass.: Harvard University Press.

Leibniz, G.W. [1701]: "Mémoire de M.G.G. Leibniz touchant son sentiment sur le calcul différentiel. Journal de Trévoux", in Leibniz [1858], p. 350.

Leibniz, G.W. [1716]: "Letter to Dangicourt, sur les monades et le calcul infinitésimal etc. (September 11)", in Leibniz [1789], vol. 3, pp. 499-502.

Leibniz, G.W. [1789]: *Opera Omnia*. (Edited by L. Dutens.) Geneva.

Leibniz, G.W. [1858]: *Mathematische Schriften*. (Volume 6 of *Die Philosophischen Scriften von G.W. Leibniz*, edited by C.I. Gerhardt.) Berlin.

Luxemburg, W.A.J. [1962]: *Nonstandard Analysis. Lectures on A. Robinson's Theory of Infinitesimals and Infinitely Large Numbers*. Pasadena: Mathematics Department, California Institute of Technology. (Second corrected edition, 1964.)

Machover, M. [1967]: "Non-Standard Analysis without Tears: An Easy Introduction to A. Robinson's Theory of Infinitesimals", Technical Report no. 27, Office of Naval Research, Information Systems Branch. Jerusalem: The Hebrew University of Jerusalem.

Meheus, J. (ed.) [2002]: *Inconsistency in Science*. Dordrecht: Kluwer Academic Publishers.

Moore, G.H. [1982]: *Zermelo's Axiom of Choice: Its Origins, Development, and Influence*. New York: Springer-Verlag.

Resnik, M. [1997]: *Mathematics as a Science of Patterns*. Oxford: Clarendon Press.

Robinson, A. [1967]: "The Metaphysics of the Calculus", in Lakatos (ed.) [1967], pp. 27-46. (Reprinted in Robinson [1979*b*], pp. 537-555.)

Robinson, A. [1974]: *Non-Standard Analysis*. (First edition, 1966.) Amsterdam: North-Holland. (Reprinted edition Princeton, NJ: Princeton University Press, 1996.)

Robinson, A. [1979*a*]: *Selected Papers of Abraham Robinson. Volume 1: Model Theory and Algebra.* (Edited by H.J. Keisler, S. Körner, W.A.J. Luxemburg and A.D. Young.) New Haven: Yale University Press.

Robinson, A. [1979*b*]: *Selected Papers of Abraham Robinson. Volume 2: Nonstandard Analysis and Philosophy*. (Edited by H.J. Keisler, S. Körner, W.A.J. Luxemburg and A.D. Young.) New Haven: Yale University Press.

Shapiro, S. [1991]: *Foundations Without Foundationalism: A Case for Second-order Logic*. Oxford: Clarendon Press.

van Fraassen, B.C. [1980]: *The Scientific Image*. Oxford: Clarendon Press.

van Heijenoort, J. (ed.) [1967]: *From Frege to Gödel.* Cambridge, Mass.: Harvard University Press.

Zermelo, E. [1904]: "Beweis, dass jede Menge wohlgeordnet werden kann (Aus einem an Herrn Hilbert gerichteten Briefe)", *Mathematische Annalen 59*, pp. 514-516. (English translation in van Heijenoort (ed.) [1967], pp. 139-141.)

Zermelo, E. [1908*a*]: "Neuer Beweis für die Möglichkeit einer Wohlordnung", *Mathematische Annalen 65*, pp. 107-128. (English translation in van Heijenoort (ed.) [1967], pp. 183-198.)

Zermelo, E. [1908*b*]: "Untersuchungen über die Grundlagen der Mengenlehre. I", *Mathematische Annalen 65*, pp. 261-281. (English translation in van Heijenoort (ed.) [1967], pp. 199-215.)

Chapter 6

MATHEMATICAL PROGRESS AS INCREASED SCOPE

Madeline Muntersbjorn
University of Toledo, Ohio

Abstract: Well-chosen languages contribute to problem-solving success in the history of mathematics. Innovations in notation may not constitute progress on their own for philosophers who reject formalism. Yet successful languages are important concomitants of progress insofar as they enable mathematicians to state claims more broadly and recognize obscure relationships between different branches of mathematics. Platonists recognize the importance of Poincare's 'happy innovations of language' as a means whereby ever more mathematical reality is revealed. However, the distinction between what mathematics is about and the formal means used to study mathematics can rarely be made precise outside of isolated historical contexts. Understanding mathematical progress as increased scope is thus an alternative to Platonism's "mathematical progress as genuine discovery" and formalism's "mathematical progress as clever invention."

Key words: mathematical progress, scope, notation

What is mathematical progress? If mathematicians make discoveries, then the applicability of mathematics to practical problems is explained easily. Mathematical discoveries model natural phenomena readily because they themselves are natural phenomena, albeit of an abstract and nonphysical kind. However, if mathematical facts simply lay dormant waiting for us to find them, it is unclear exactly where they are and why we find some facts, rather than others, at any given time. G. H. Hardy (1929) compares mathematicians to surveyors:

> I have myself always thought of a mathematician as in the first instance an *observer*, a man who gazes at a distant range of mountains and notes down his observations. His object is simply to distinguish clearly and notify to others as many different peaks as he can. There are some peaks

B. Van Kerkhove and J.P. van Bendegem (eds.), Perspectives on Mathematical Practices, 107–117.

> which he can distinguish easily, while others are less clear. He sees A sharply, while of B he can obtain only transitory glimpses. ... In other cases perhaps he can distinguish a ridge which vanishes in the distance, and conjectures that it leads to a peak in the clouds or below the horizon. But when he sees a peak he believes that it is there simply because he sees it. If he wishes someone else to see it, he *points to it*.... (P. 18.)

Even if this image captures what it feels like to do mathematics, many philosophers insist that the mathematician who claims that she perceives something real when she does mathematics is simply wrong. The "access problem" confronting Platonism may be restated: towards what are these surveyors pointing? In particular, what are the tools of the trade mathematicians use to map ever more extensive territory? Is it plausible that this territory remains wholly unaffected by these activities? While I am *sympathetic* to mathematical realism, the passivity inherent in this image does not strike me as *tenable*, for it fails to capture the active engagement characteristic of mathematical practices.[1] As Hardy notes, the image, when pressed, suggests that mathematical proofs are naught but "rhetorical flourishes designed to affect psychology" (ibid., 18). For my part, this consequence threatens to undermine, rather than support, the realist conviction that mathematicians actively engage a mind-independent reality via their practices.

If mathematicians construct their subject matter, the variety of mathematical practices is explained easily. However, if mathematics is just an elaborate invention, the applicability of mathematics, as well as its ability to transcend cultural boundaries, becomes an enigma. Some see mathematical practice as an elaborate system of cultural conventions, constrained, if at all, by our cognitive endowments. But this account frustrates those who see it as inadequate to the task of explaining mathematical success. As one of my students once said, "I don't think we are smart enough to invent something that powerful. Something else must be going on." This perspective may be dismissed as an unwarranted lack of species-esteem. Yet, even those who celebrate the awesome power of the collective human mind must acknowledge the unrivaled unity of mathematics as well as the genuine struggle of mathematicians who try to capture this unity formally. Ancient Babylonian mathematics differs from Ancient Greek mathematics, but insofar as they are both examples of *mathematics*, they share a conceptual core in common. For example, there are many ways of expressing the singular truth that most frequently goes by

1 "One may divide philosophies into *sympathetic* and *unsympathetic*, those in which we should like to believe and those which we instinctively hate, and into *tenable* and *untenable*, those in which it is possible to believe and those in which it is not" (ibid., 3).

the arbitrary name, "The Pythagorean Theorem." The more vigorously we pursue new and improved articulations of the conceptual unity of mathematics, the more powerful mathematics becomes, despite the fact that individual articulation efforts meet with stunning failures. The notorious difficulties that plague foundational approaches to mathematics may be seen as evidence that the "common core" is a cultural myth. But this argument can be turned inside out: If mathematics is completely under our control, why does it so frequently happen that mathematical projects fail to turn out the way we expect them to? Why are we so frequently surprised and disappointed when promising programs fail? My student's view may be rephrased, "I don't think we are dimwitted enough to invent something that recalcitrant. Something else must be going on."

A sympathetic and tenable account of mathematical progress will reconcile the cultural diversity of mathematical practices with the conceptual unity of mathematical phenomena, without dismissing either as an illusion. While I do not have such an account in hand, I do have some suggestions about the kinds of research into the history and philosophy of mathematics we can do in an effort to develop such an account. Like the Perspectives on Mathematical Practice conference, this research programme is inspired by the work of Imre Lakatos. In particular, it is inspired by Lakatos' views on the relation between mathematical rigor and mathematical content. Consider the following question:

For Lakatos, increases in mathematical rigor lead to __________ in mathematical content.

(a) increases (b) decreases

Rigor and content, or to use Hardy's diction again, *purification* and *enlargement*, come together in an inconclusive exegetical question (op cit., 13). Consider these passages from *Proofs and Refutations*:

> Omega: …proof-analysis, when increasing certainty, decreases content. Each new lemma in the proof-analysis, each corresponding new condition in the theorem, reduces its domain. … We should have a counterweight against the content-decreasing pressure of rigor. (P. 57.)

> Certainty is never achieved'; 'foundations' are never found—but the 'cunning of reason' turns each increase in rigour into an increase in content, in the scope of mathematics. (P. 56.)

Omega worries that an emphasis on rigor leads to an impoverishment of mathematical content. In the footnote, Lakatos tempers this worry by suggesting that, even when we try and "squeeze the content out" of our proofs in an effort to be more rigorous, mathematical reasoning being what it is, we end up with richer content and a broader context despite ourselves.

Lakatos does not spell out in detail how the quest for purity leads to an enlargement of mathematics: "this story is beyond our present investigation" (ibid.). His disputants discuss the relation between rigor and content, but fail to achieve consensus. Indeed, maybe there isn't a single right answer to the fill-in-the-blank question—not simply because Lakatos died before articulating his views on the matter—but because he did not think that an account could be given. Aphorisms such as, "The growth of knowledge cannot be modeled in any given language" and "Science teaches us not to respect any given conceptual-linguistic framework lest it should turn into a conceptual prison," suggest that we are destined to debate the question in perpetuity (93).

However, the best answer to the question is (a). Increases in rigor lead to increases in content as the scope of mathematics increases. Recall Fermat's famous marginalia. Perhaps Fermat did not really have a proof of his theorem in hand any more than Lakatos had a finished story to tell about rigor and content. However, both could see *what* must be the case before they were able to explain *why*. The proposed research programme involves telling stories designed to explain the connections between rigor, content and scope. Good stories will be detailed narratives that respect the historical record and help reconcile the diversity of mathematical practices with the unity of mathematical phenomena. My research into the development of mathematics in the 17th C. has led me to formulate the following plot-summary regarding the relation between rigor, content and scope. Increases in scope are achieved when one generation's *form* becomes another generation's *content*. Frequently, novel notation is introduced to abbreviate linguistic patterns of reasoning or implicit mental operations. In synchronic problem-solving domains, we regard any given mathematical sign as "an arbitrary mark, with a fixed interpretation," as Boole correctly notes in his *Laws of Thought* (25). Boole also notes, "That language is an instrument of human reason, and not merely a medium for the expression of thought, is a truth generally admitted" (24). However generally admitted this distinction may have been in the 19th C., this distinction continues to deserves closer scrutiny. In particular, we should pay closer attention to the phenomenon whereby arbitrary but convenient abbreviations may become laden with novel content as the scope of mathematics increases. Sometimes, mathematicians are compelled to expand beyond the "fixed" interpretation originally assigned to mathematical notations in diachronic explanatory contexts. As a result, informal strategies of reasoning and tacit insights are transformed into formal languages capable of conveying explicit content. Thus, examining mathematical representational practices over time should yield insight into the nature of mathematical progress.

Hardy expects a philosopher of mathematics to explain, rather than dismiss, the "invincible feeling that, if Littlewood and I both believe Goldbach's theorem, then there is something, and that the same something, in which we both believe…" (24). In order to meet this expectation, I must take issue with the remainder of his sentence: "…and that that same something will remain the same something when each of us is dead and when succeeding generations of more skilful mathematicians have proved our belief to be right or wrong" (ibid.). The death of a particular mathematician, in all likelihood, cannot substantially alter a conjecture. However, a proof of a conjecture, in all likelihood, can and does substantially alter it. This alteration comes about not simply because the conjecture may then be upgraded to theorem status, but because proofs have the power to reveal previously hidden relationships which, once explicit, extend the scope of the theorem and, thereby, change what the theorem is about.

Several astute observers of mathematical practice have noticed the phenomenon, wherein a sign introduced as a convenient heuristic is subsequently taken to refer to features of mathematical reality. Judgments concerning the significance of such developments vary. For Shapiro (2000), this phenomenon tempts unwary mathematicians into formalism:

> Throughout history, mathematicians have had occasion to introduce symbols which, at the time, seemed to have no clear interpretation. … Although the newly introduced 'entities' proved useful for applications within mathematics and science, in their philosophical moments some mathematicians did not know what to make of them. What are imaginary numbers, really? A common response to such dilemmas is to retreat to formalism. (P. 141.)

My proposed research programme should not be confused with formalism. Shapiro correctly notes that our philosophical interests are not served by retreating from the metaphysical messiness of this phenomenon by denying the existence of mathematical entities. Instead, we should ask, what kinds of things are mathematical entities? They are not mere signs. They are the kinds of things about which one can construct natural histories, or *Naturgeschicten*, as Wittgenstein suggests. However, these histories will be illuminating only if they feature changes in mathematical signification as central parts of the plot, as I have argued elsewhere (see, e.g., Muntersbjorn 2003).

As Shapiro notes, Frege was an exceptional mathematician who noticed this phenomenon without retreating into formalism. Two of Frege's philosophical goals were to block the retreat to formalism and to prevent future mathematicians from making new metaphysical messes. At the end of the introduction to the first volume of his *Grundgesetze*, he writes:

> It is strange how an imprecise spoken or written form of expression, which may originally only have been used for reasons of convenience and brevity, yet with full awareness of its imprecision, can eventually confuse thought, when that awareness has faded. Numerals end up being taken for numbers, names for what is named, mere notation for the real object of arithmetic. Such experiences teach us how necessary it is to place the strictest demands on precision of spoken and written forms of expression. (1997, pp. 210-211.)

In short, Frege sees this phenomenon not as paradigmatic of the growth of mathematical knowledge, but as a mistake the astute avoid making. Similarly, Hardy remarks that, in the course of our mathematical investigations,

> We may even find it necessary to guide our thoughts by the introduction of new formalism, and it is quite likely that, if we do, we shall use over again the same symbols that we have used already. And here, of course, lies a danger; for we may be tempted to forget that we are using the same symbols in different contexts with different aims; even Russell has been accused of making this mistake by logicians of the more formal schools. (Op cit., 17.)

However, not everyone sees the same danger inherent in either the introduction of new formalism or the re-interpretation of existing symbolism. Whitehead (1911) portrays this process in a more positive light by referring to one of the more briefly recounted examples:

> The idea of zero probably took shape gradually from a desire to assimilate the meaning of this mark to that of the marks, 1, 2, … 9, which do represent cardinal number. This would not represent the only case in which a subtle idea has been introduced into mathematics by a symbolism which in its origin was dictated by practical convenience. (P. 64).

Langer (1953) also cites "0" as an example in support of her contention that,

> There is something uncanny about the power of a happily chosen ideographic language; for it often allows one to express relations which have no names in natural language and have therefore never been noticed by anyone. Symbolism, then, becomes an organ of discovery rather than mere notation. (P. 60).

The elasticity of mathematical language, or the ability of mathematical expressions to be stretched beyond their original meanings, was noted by Poincaré (and subsequently cited by Lakatos). Unlike Frege and Hardy, Poincaré sees this phenomenon as a mark of mathematical significance. In the second chapter of his *Science and Method*, "The Future of Mathematics," Poincaré (1929) writes:

> [M]athematics is the art of giving the same name to different things. It is proper that these things, differing in matter, be alike in form, so that they may, so to speak, run in the same mold. When the language is well chosen, we are astonished to learn that all the proofs made for a certain object apply immediately to many new objects; there is nothing to change, not even the words, since the names have become the same.
>
> A well-chosen word usually suffices to do away with the exceptions from which the rules stated in the old way suffer; this is why we have created negative quantities, imaginaries, points at infinity and what not. And exceptions, we must not forget, are pernicious because they hide the laws.
>
> Well, this is one of the characteristics by which we recognize the facts which yield great results. They are those which allow of these happy innovations of language. The crude fact then is often of no great interest; we may point it out many times without having rendered great service to science. It takes value only when a wiser thinker perceives the relation for which it stands, and symbolizes it by a word. (P. 375.)

I cannot do justice to Poincaré's philosophy of mathematics in the remainder of this essay. Instead, I make four brief observations, inspired by this passage and offered in support of my contention that the elasticity of mathematical language is a poorly understood virtue rather than a lamentable vice.

First, Poincaré's view helps explicate the historian's judgment that we can use the presence or absence of a well-chosen language as a non-arbitrary way to distinguish between who did and did not make a particular mathematical discovery. As Baron (1969) notes,

> Mathematical invention is a process of continuous change and development rather than something which takes place at a given point in time, but if it be considered necessary to draw a line between those mathematicians of the seventeenth century who "had the calculus" and those who "had not" the line would inevitably exclude Barrow on the

grounds that he exhibited no calcular rules and used no specialized notation or symbolism. (Pp. 251-252.)

Thus, Leibniz had the calculus but Fermat did not. Fermat was certainly aware of the "crude fact" that there was a mathematical relation between the solution to tangent problems and the quadrature of particular curves. The unique and original diagrams in his "Treatise on Rectification" exhibit instances of this particular relation in geometric form. Nevertheless, this particular fact only took on value in light of Leibniz' "happy innovations of language," namely the introduction of specialized symbols for the operations of integration and differentiation. Fermat's methods were less successful, not only because they were less "rigorous" than Leibniz', but also because they were less widely circulated and hence less accessible. Failure to achieve a broad audience cannot be distinguished sharply from failure to achieve a method of increased scope and rigor, a point I return to below.

Second, Poincaré is often described as a conventionalist. Sentences like, "this is why we have created negative quantities, imaginaries, points at infinity and what not," taken in isolation, certainly seem to identify mathematical objects as arbitrary artifacts. Yet, the next sentence doesn't fit with this interpretation. "Exceptions are pernicious because they hide the laws" suggests that our creations are intended to reveal the laws governing mathematical reality, laws we do not author but are bound to obey. "Well-chosen words" are the key to unlocking the apparent tension between creation and discovery. We state rules, but sometimes exceptions to these rules emerge because of infelicities in our expressions. So we develop better languages to restate the rules so that they more clearly reveal the relations governed by the rules, i.e., the laws. Perhaps what Poincaré should have written was, this is why we have created "negative quantities," "imaginaries" and what not. Writing before the well-chosen notational practice of placing quotes around names when they are mentioned but not used, presumably Poincaré expected his reader to understand that names, not objects, are deliberately created in response to perceived relations between salient facts.

Third, Poincaré writes that "Invention is discernment, choice," but not choice in the sense that a diner in a cafeteria chooses the chicken instead of the fish. Mathematical choices are not made from given and immediately perceptible arrays of options. In fact, for Poincaré, mathematical choices are not necessarily conscious! The conscious self and the subliminal self collaborate to reveal esthetically satisfying mathematical harmonies. Seeing these harmonies clearly necessitates taking the pains to make them explicit in our discourse. Not surprisingly, he's a bit fuzzy on the details of this process. However, he clearly did not share Frege's view that we must rid the philosophy of mathematics from such overtly psychological concepts as "subliminal selves," "unconscious choices," and "esthetic satisfaction." The proposed research programme of investigating how *implicit* patterns of reasoning are given *explicit* names in formal languages similarly blurs the distinction between philosophy and psychology of mathematics.

Finally, Poincaré's extract challenges Lakatos' distinction that, "Heuristics is concerned with language-dynamics, while logic is concerned with language-statics" (op cit., 93). This distinction seems apropos in light of Frege's call to fix the logical uses of language. However, as Poincaré reminds us, new discoveries frequently depend on the articulation of proofs. A proof designed with one purpose in mind may end up serving other purposes as well. In other words, logicians overtly concerned with "statics" frequently invent new languages, compelled, as Frege claimed to be, by "a necessity inherent in the subject matter" (van Heijenoort 1970, 7). They may, thereby, contribute to "dynamics," or changes in mathematical languages. As McLarty (2000) points out, the "anti private-rigor" argument implies that confronting logical gaps and cultivating shared means of expression are not separate endeavors:

> Progress towards rigor includes finding and filling in gaps in definitions and proofs. That's what we most often look at. But the other aspect is replacing idiosyncrasy with widely shared means of expression. These are not separate parts—as if you could do one without the other—but inseparable aspects of one effort. (P. 270.)

Frege's motivation for developing his *Begriffsschrift* was to "break the domination of the word over the human spirit"(ibid.). It is not clear whether Frege achieved this goal or whether anyone ever could. However, his concept-script clearly sparked a revolution in mathematical languages. These newly developed languages made it possible for us to explore more mathematical terrain than before. As a result, Frege's project, which was to increase rigor, also led to an increase in content.

Despite his distrust of language, Frege wielded it with style. In particular, Frege was a master author of illuminating metaphors. The *Begriffsschrift* offers a similar irony. In order to fix mathematical foundations, Frege felt compelled to invent a new mathematical language—yet this new mathematical language changed mathematics such that today we doubt whether its foundations can be fixed. Perhaps things did not turn out the way he had hoped, but Frege contributed to mathematical progress nonetheless by broadening our mathematical scope. Because of the pains Frege took to say what he alone could see, the mathematical community as a whole can now see more. Frege himself hints at this possibility in one of his finest metaphors, wherein he compares innovations in logical languages to improvements in optical instruments:

> I believe that I can best make the relation of my ideography to ordinary language clear if I compare it to that which the microscope has to the eye. Because of the range of its possible uses and the versatility with which it can adapt to the most diverse circumstances, the eye is far superior to the microscope... But, as soon as scientific goals demand great sharpness of resolution, the eye proves insufficient. The microscope, on the other hand, is perfectly suited to precisely such goals, but that is just why it is useless for all others.
>
> This ideography, likewise, is a device invented for certain scientific purposes, and one must not condemn it because it is not suited to others. If it answers to these purposes in some degree, one should not mind the fact that there are no new truths in my work. (Ibid., 6.)

Frege's illuminating metaphor reappears as an aphorism in Langer (1953): "Logic is to the philosopher what the telescope is to the astronomer: an instrument of vision" (41). However, Frege's purely instrumentalist defense of his *Begriffsschrift* is weaker than need be. He imagines, as many authors of new mathematical languages do, that his concept-script is simply a collection of new names for extant things. And while this may have been true at the time of their formulation, *when the implicit becomes explicit the boundary between form and content changes*. Well-chosen mathematical languages contribute to our problem-solving success. The success of certain problem solving techniques, especially those involving new representational strategies, can be explained via an appeal to more abstract objects. In these cases, the referents of these symbols become "objects of explanation in terms of theories of a higher level describing a further and deeper layer of reality," to paraphrase Popper. (Cf. the PMP paper of Eduard Glas wherein he urges the relevance of Popper's views for the philosophy of mathematics, see chapter 3 in this volume.) Thus while Frege did not set out to develop signs for "new truths," the fact that he did develop signs for what were previously tacit patterns of reasoning means that his descendants are able to discern new truths.

But are these truths genuinely new? A Platonist, inspired by Hardy's image, may insist that representational innovations simply disperse clouds that previously obscured our vision of distant peaks that were there all along. But the more closely we examine the relation between mathematical expressions and mathematical entities—between the means of cloud dispersal and the peaks themselves—the more elusive the distinction between the two becomes. Within the context of a historical narrative, with a clearly defined before and after, we can formulate a context-dependent distinction between form and content and avoid the retreat to formalism.

But the distinction between what mathematics is about and the formal strategies used to investigate what mathematics is about cannot be made precise in the abstract without placing indefensible borders around mathematical reality. Hardy presents a brief account of the "presuppositions and prejudices with which a working mathematician is likely to approach philosophical or logical systems" (op cit. 5). In particular, "Mathematicians have always resented attempts by philosophers or logicians to lay down dogmas imposing limitations on mathematical truth or thought" (ibid.). However, the only way to secure an open future for mathematics while at the same time respecting the intuition that, "In *some* sense, mathematical truth is part of objective reality," is to reject the prejudice that mathematical truths are "immutable," "unconditional," "absolute and independent of our knowledge of them" (ibid., 4). To return to the image with which this essay began, mathematicians are not like mapmakers who adhere to the environmentalist's ethic, "take only memories—leave only footprints." They are more like *terraformers*, science-fiction engineers who travel to inhospitable planets and struggle to make alien landscapes suitable for human settlement by adapting them to our perceptual needs and abilities via innovations in formal systems of signification.

REFERENCES

Baron, Margaret. (1969). The Origins of the Infinitesimal Calculus. Dover.
Boole, George. ([1854] 1958). An Investigation of the Laws of Thought. Dover.
Frege, Gottlob. (1997). The Frege Reader, Michael Beaney, ed. & trans. Blackwell.
Hardy, G. H. (1929). "Mathematical Proof," Mind **38**: 1-25.
Lakatos, Imre. (1976). Proofs and Refutations. Cambridge.
Langer, Susanne. (1953). An Introduction to Symbolic Logic. Dover.
McLarty, Colin. (2000). "Voir-Dire in the Case of Mathematical Progress," E. Grosholz & H. Breger, eds., The Growth of Mathematical Knowledge. Kluwer: 269-280.
Muntersbjorn, Madeline. (2003). "Representational Innovation and Mathematical Ontology," Synthese **134**: 159-180.
Shapiro, Stewart. (2000). Thinking About Mathematics. Oxford.
Van Heijenoort, Jean. (1970). Frege and Gödel. Harvard.
Whitehead, A. N. (1911). An Introduction to Mathematics. New York: Henry Holt & Co.

Chapter 7

PROOF IN C17 ALGEBRA

Brendan Larvor
University of Hertfordshire, UK

Abstract: This paper considers the birth of algebraic proof by looking at the works of Cardano, Viète, Harriot and Pell. The transition from geometric to algebraic proof was mediated by appeals to the Eudoxan theory of proportions in book V of Euclid. The crucial notational innovation was the development of brackets. By the middle of the seventeenth century, geometric proof was unsustainable as the sole standard of rigour because mathematicians had developed such a number and range of techniques that could not be justified in geometric terms.

Key words: Algebra, proof, Cardano, Viète, Harriot, Pell

By the middle of the sixteenth century there was in Europe, on the one hand, geometry, which had well-established standards and methods of proof, and a large body of actual proofs. On the other hand, there was an emerging body of analytic techniques that did not have their own criteria or means of proof. These techniques developed naturally out of simple recipes for performing arithmetical calculations such as the rule of three or the various methods of long division. Having established techniques for finding square roots, it was natural for arithmeticians to extend these techniques to problems that we would nowadays express in quadratic equations (we still speak of the 'roots' of an equation). In short, geometers looked for theorems with proofs, but people doing what came to be called 'algebra' or 'specious analysis' were looking for solutions. This division is reflected in the titles of algebra books such as Cardano's *Ars Magna* or Harriot's *Artis Analyticae Praxis*. While the word '*ars*' (art in the unromantic sense of craft or technique) was often used, the word '*scientia*' is pointedly absent from these book titles. '*Scientia*' was a highly contested term but the principal source of its meaning was still Aristotle's *Posterior Analytics*, in which the title of

B. Van Kerkhove and J.P. van Bendegem (eds.), Perspectives on Mathematical Practices, 119–133.

science was reserved for systematic, deductive knowledge. In the sixteenth century, geometry was widely[1] taken to fulfil this requirement while algebra did not. However, by the middle of the seventeenth century we that find that algebra is able to offer proofs in its own right. That is, by that time algebraic argument had achieved the status of proof. How did this transformation come about?

This question could easily occupy an entire monograph. An apparently continuous historical narrative that registered the many small steps on the journey, and that paid due attention to the unevenness of the development (for fear of presenting an excessively streamlined and whiggish account) might need more than one book. Historical development is always uneven, as there are always individuals ahead of their times and a greater number behind. The development of early modern algebra is especially uneven due to the diversity of sources: to an indigenous European tradition of reckoning were added rediscovered Diophantus and the works of Islamic mathematicians. Moreover, it was at about this time that national styles and rivalries started to become pronounced in European science. In view of these complexities, the most we can hope to achieve in a short paper is a comparison of snapshots taken at significant moments in the story. This paper offers a small collection of still photographs rather than the sort of cinematic sequence that gives an illusion of continuity.

Before we open the photo-album, permit me a methodological note. The initial sketch of a mathematics divided between problem-solving 'analysis' (arithmetic-cum-algebra) and theorem-proving geometry is broadly-speaking right. Indeed, some Renaissance mathematicians regretted this divide and dreamed of a unified, 'universal' mathematics ('*mathesis universalis*').[2] However, to set up this division as an absolute distinction is to make a mystery of the fact that it was eventually overcome. Absolute distinctions create insoluble historical problems (as Kuhn discovered). Instead, we

[1] But not universally. See (Mancosu 1992). This paper considers the debate following the publication in 1547 of the *Commentarium de certitudine mathematicarum disciplinarum* by Alessandro Piccolomini (1508-1578). According to Mancosu, Piccolomini attempted to refute "a widespread argument which aimed at showing the certitude of mathematics (asserted by Aristotle and reiterated by Averroes and a long list of Aristotelian commentators) arguing from the assumption that mathematics makes use of the highest type of syllogistic demonstrations" (p. 244). The chief point stands, since the mathematics to which this controversy refers is that of Euclid. As Mancosu put it, the question was, "What is the relationship between Aristotelian logic and Euclidean mathematics?" (p. 242). For Piccolomini and his fellow renaissance writers, the question did not arise with respect to the analytic art. Things were quite different a century later, when Hobbes, Wallis and Barrow took up the question of mathematical certainty afresh. Then, the status of algebra was at the centre of the dispute. See (Sasaki 1985)

[2] See (Sasaki 1985) pp. 91-92; (Mancosu 1996) p. 86.

should recall that to draw a distinction is at the same time to make a connection. That is, we must look for points in the historical record where the apparently absolute distinction breaks down: in other words, we look for intimations of proof and rigour on the un-rigorous, 'heuristic' side of the division: in the analytic practice, in the modes of argument, but also in the structure of the books themselves and the language used.

1. GIROLAMO CARDANO (1501-1576)

The subject of our first snapshot is Girolamo Cardano's *The Great Art* (1545), in which he systematises and proves the thirteen solutions by radicals of the cubic. There are thirteen solutions because his proofs are geometric. He interprets the unknown as a line segment and the coefficients as lines, areas or volumes so as to preserve the homogeneity of the sum. ('Homogeneity' means that volumes are only added to volumes, areas to areas and so forth. To achieve this with a cubic equation, Cardano read the coefficient of the squared term as a line and that of the linear term as an area, while he treated the constant term as a volume. Then, the whole equation is in volumes.) Consequently, he could only recognise positive coefficients (for what sense can be given to negative areas or volumes?), so rather than just *the* cubic he had thirteen variations. He had no symbolism (except the use of letters to label points on diagrams) and consequently none of the notational machinery nowadays associated with algebra. Rather, he wrote everything out in abbreviated prose and gave his proofs in the Euclidean style (see figure 1). This geometrical rigour constrained the scope of his algebra. For him there could be no rigorous treatment of equations of degree greater than three because "nature will not allow it"[3]. That is, there are just three dimensions in space, and geometry is the means of proof, so nothing can be proved about equations of degree greater than three.

This geometric standard of rigour was a source of difficulty for Cardano because in the appendix to *Ars Magna* he explains how to solve a certain class of bi-quadratics (this result was due to his student, Ferrari). What is the status of Ferrari's argument to show that his solutions of these bi-quadratics are correct? It cannot be proof (by Cardano's standards), yet it is entirely persuasive. This, though, is the least of Cardano's difficulties. Using modern

[3] "For as *positio* refers to a line, *quadratum* to a surface, and *cubus* to a solid body, it would be very foolish for us to go beyond this point, nature does not permit it." (Witmer and Cardano 1968) p. 9. "Nanque cum positio lineam, quadratum superficiam, cubus corpus solidum referat, nae utique stultum fuerit, nos ultra progredi, quo naturae no licet." (Cardano 1570) p. 6.

notation, consider his solution by radicals of the 'irreducible' case of a cubic with three real roots. If $\boldsymbol{a}$ and $\boldsymbol{N}$ are positive and

$$x^3 = ax + N \text{, then:}$$

$$x = \sqrt[3]{\frac{N}{2} + \sqrt{\left(\frac{N}{2}\right)^2 - \left(\frac{a}{3}\right)^3}} + \sqrt[3]{\frac{N}{2} - \sqrt{\left(\frac{N}{2}\right)^2 - \left(\frac{a}{3}\right)^3}}$$

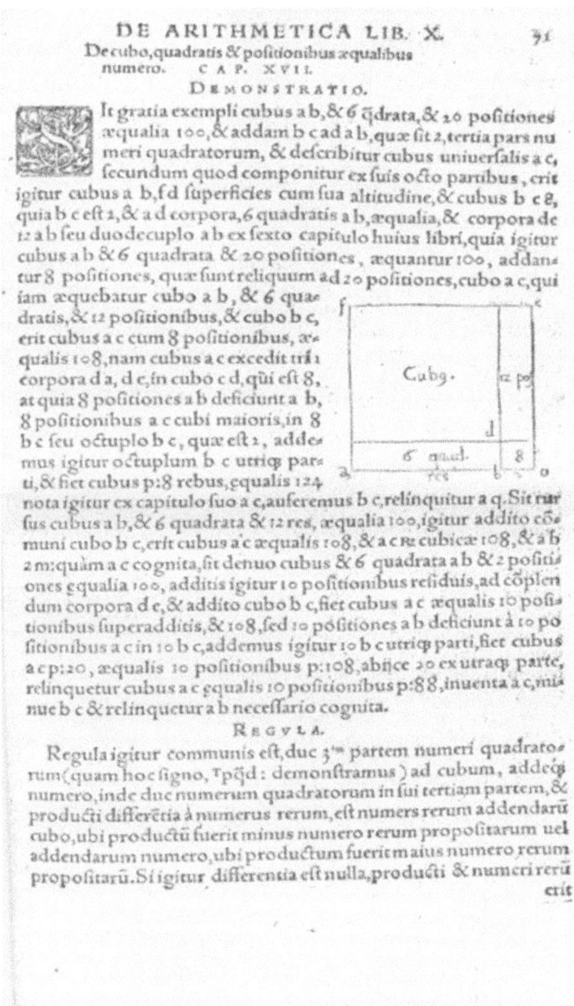

DE ARITHMETICA LIB. X. 71

De cubo, quadratis & positionibus æqualibus numero. CAP. XVII.

DEMONSTRATIO.

Sit gratia exempli cubus a b, & 6 q̃drata, & 20 positiones æqualia 100, & addam b c ad a b, quæ sit 2, tertia pars numeri quadratorum, & describitur cubus uniuersalis a c, secundum quod componitur ex suis octo partibus, erit igitur cubus a b, f d superficies cum sua altitudine, & cubus b c 8, quia b c est 2, & a d corpora, 6 quadratis a b, æqualia, & corpora d e 12 a b seu duodecuplo a b ex sexto capitulo huius libri, quia igitur cubus a b & 6 quadrata & 20 positiones, æquantur 100, addantur 8 positiones, quæ sunt reliquum ad 20 positiones, cubo a c, qui iam æquebatur cubo a b, & 6 quadratis, & 12 positionibus, & cubo b c, erit cubus a c cum 8 positionibus, æqualis 108, nam cubus a c excedit tria corpora d a, d c, in cubo c d, qui est 8, at quia 8 positiones a b deficiunt a b, 8 positionibus a c cubi maioris, in 8 b c seu octuplo b c, quæ est 2, addemus igitur octuplum b c utrique parti, & fiet cubus p:8 rebus, equalis 124 nota igitur ex capitulo suo a c, auferemus b c, relinquitur a q. Sit rursus cubus a b, & 6 quadrata & 12 res, æqualia 100, igitur addito communi cubo b c, erit cubus a c æqualis 108, & a c R cubicæ 108, & a b 2 m: quàm a c cognita, sit denuo cubus & 6 quadrata a b & 2 positiones equalia 100, additis igitur 10 positionibus residuis, ad complendum corpora d e, & addito cubo b c, fiet cubus a c æqualis 10 positionibus superadditis, & 108, sed 10 positiones a b deficiunt à 10 positionibus a c in 10 b c, addemus igitur 10 b c utrique parti, fiet cubus a c p:20, æqualis 10 positionibus p:108, abijce 20 ex utraque parte, relinquetur cubus a c equalis 10 positionibus p:88, inuenta a c, minue b c & relinquetur a b necessario cognita.

REGVLA.

Regula igitur communis est, duc 3^am partem numeri quadratorum (quam hoc signo, Tp q̃d: demonstramus) ad cubum, addeque numero, inde duc numerum quadratorum in sui tertiam partem, & producti differẽtia à numerus rerum, est numers rerum addendarũ cubo, ubi productũ fuerit minus numero rerum propositarum uel addendarum numero, ubi productum fuerit maius numero rerum propositarũ. Si igitur differentia est nulla, producti & numeri rerũ erit

Figure 1. (Cardano 1570) p. 71

Obviously when $(N/2)^2 < (a/3)^3$ the square roots are imaginary and the whole expression is complex. Nevertheless, the solutions are real. This is an early instance of a detour through the complex universe prompted by a question that makes no reference to complex numbers in either its statement or its solution.

This problem was more serious than that posed by equations with complex roots, since in those cases Cardano could simply declare such roots impossible or nonsensical (as many mathematicians did). Cubics with three real roots cannot be so dismissed. Thus there was, for Cardano, a dilemma. On one hand, he had established standards of proof, on the other he had persuasive results that were impossible in principle to prove within these standards. As we shall see, this dilemma intensified in the succeeding century. What was, for Cardano, a small trickle of results that could not be modelled in the geometry of solids with finite magnitudes, broadened to a flood. In retrospect, we may (in a Popperian spirit) regard Cardano's irreducible cubics and Ferrari's bi-quadratics as 'refutations' of the metaphysics and methodology implicit in Cardano's geometric standards of rigour. However, we should note, first, that Cardano had no such perspective. Second, his anomalous cases might have remained as no more than recondite unsolved problems, and posed little threat to the established methodological order, had they not been but the first of many.

2. FRANÇOIS VIÈTE (1540-1603)

Viète's principal achievement is the introduction of a recognisably modern symbolism for variables, coefficients and some operations (though he still wrote everything else out longhand; note also the use of double lines to indicate subtraction). Nevertheless, Viète's algebra seems firmly within the tradition of problem-solving technique since he gives little or nothing in the way of proof. Consider, for example, Proposition XVI of his *Ad Logisticen Speciosam Notæ Priores* (Vietae 1646) p. 20, which is in fact a problem-to-solve rather than a hypothesis-to-prove: "To subtract the cube of the difference between two roots from thc cube of their sum." The argument consists of a single sentence: "Let the individual solids making up the cube of $A-B$ be subtracted from the individual solids making up the cube of $A+B$." This is, strictly speaking, a procedure rather than a proof, and he carries it out in the next sentence to get the answer: in modern notation, $6A^2B+2B^3$ (see figure 2). However, we cannot conclude from the brevity of this discussion that Viète was indifferent to proof. Certainly, he hoped to illustrate the expressive power and heuristic efficiency of his symbolism. However, his texts have the form (if not the substance) of a deductive, 'scientific' system. He labels his problems as 'propositions' and his solutions as 'theorems', and he makes reference to Aristotle's *Posterior Analytics*.[4] In particular he claims that theorems proved by his art conform to the laws governing the relation of attribute to subject, laid down in *Posterior Analytics*, Book I, Part IV. So in terms of the *Commentarium* that Mancosu discusses, Viète claims for algebra (specifically, for his 'zetetics') the status attributed to Euclidean mathematics by Aristotelian tradition.

The fact that Viète could present a calculation with letters standing in for arbitrary quantities as a proof, however modest, indicates a change in the conception of number itself. Viète did not interpret numbers geometrically as Cardano did, even though he retained the old geometric vocabulary of 'squares' and 'cubes' (as indeed we do today). That is why he felt no obligation to locate a 'line' in a diagram (it makes no sense, to use the example in hand, to ask about the relative positions of *A* and *B*). By contrast, each of Cardano's proofs implies a diagram: the elements of the proof are elements of the diagram (line-segments, areas and volumes with determinate relative positions). The diagram is essential—it supplies the terms of the theorem with their meanings (though in fact not every proof in *Ars Magna* has a diagram printed with it, since the proofs are all rather similar, so having seen one diagram it is easy for the reader to supply the others). None of this holds for Viète, whose few diagrams appear late in the text, after his algebra is established. Another indication of this change is the fact that Viète was not bound to three dimensions, even in principle. Cubes of cubes make sense in Viète's mathematics. Cardano did work with negative and complex numbers, and did contemplate powers higher than three, but only when reckoning the solutions to problems. The change in the mode of mathematical reasoning between Cardano and Viète indicates a change in the conception of number precisely because it takes place in the context of proof, where philosophical niceties matter.

Viète's scientific aspirations for his algebra are clearest in the programmatic part of his work. At the start of chapter two of his *Introduction to the Analytic Art* (1591), he says, "Analysis accepts as proven the well-known fundamental rules of equations and proportions that are given in the Elements". There follows a list of 'rules', the first six of which are adapted from the common notions of Euclid. The remaining ten rules are adapted from book V of Euclid (sometimes attributed to Eudoxus of Cnidus), on ratio and proportion. He then claims that, "a proportion may be

[4] See (Viete and Witmer 1983) p. 28.

said to be that from which an equation is composed and an equation that into which a proportion resolves itself."[5]

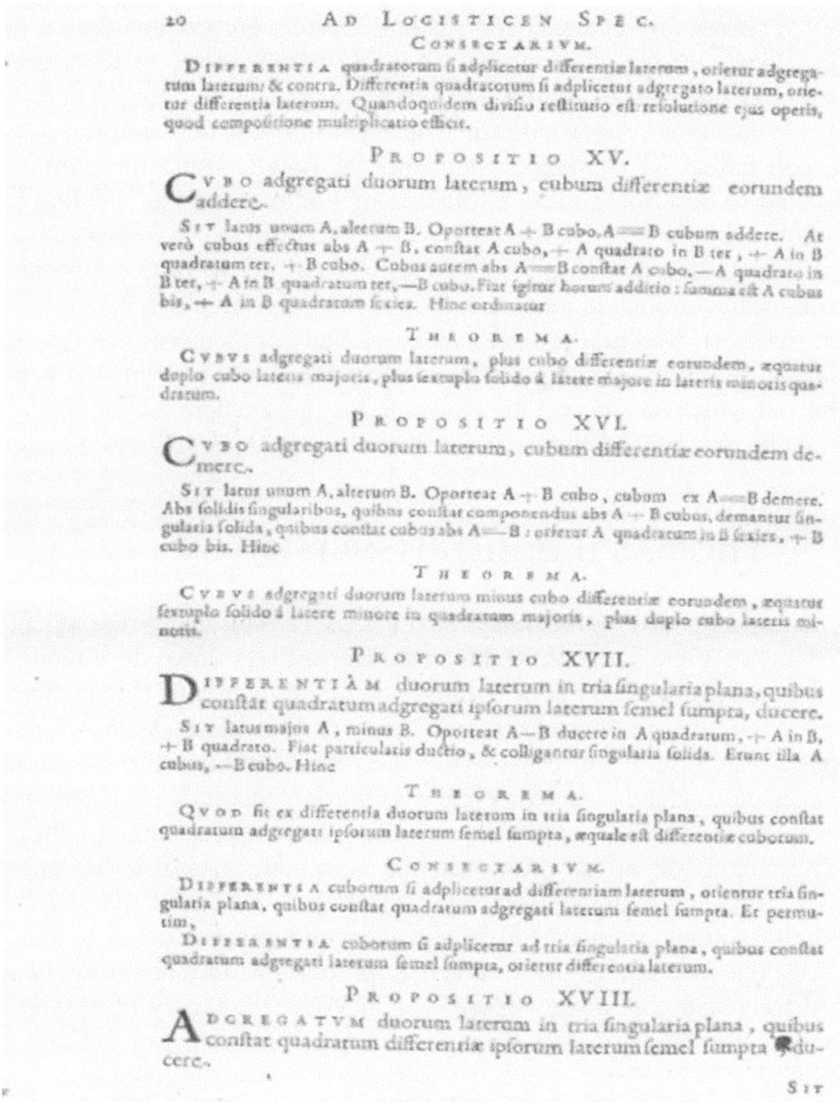

20 AD LOGISTICEN SPEC.

CONSECTARIUM.

DIFFERENTIA quadratorum ſi adplicetur differentiæ laterum, orietur adgregatum laterum: & contra. Differentia quadratorum ſi adplicetur adgregato laterum, orietur differentia laterum. Quandoquidem diviſio reſtitutio eſt reſolutione ejus operis, quod compoſitione multiplicatio efficit.

PROPOSITIO XV.

CUBO adgregati duorum laterum, cubum differentiæ eorundem addere.

SIT latus unum A, alterum B. Oporteat A + B cubo, A == B cubum addere. At verò cubus effectus abs A + B, conſtat A cubo, + A quadrato in B ter, + A in B quadratum ter, + B cubo. Cubus autem abs A == B conſtat A cubo, — A quadrato in B ter, + A in B quadratum ter, — B cubo. Fiat igitur horum additio: ſumma eſt A cubus bis, + A in B quadratum ſecies. Hinc ordinatur

THEOREMA.

CUBUS adgregati duorum laterum, plus cubo differentiæ eorundem, æquatur duplo cubo lateris majoris, plus ſextuplo ſolido à latere majore in lateris minoris quadratum.

PROPOSITIO XVI.

CUBO adgregati duorum laterum, cubum differentiæ eorundem demere.

SIT latus unum A, alterum B. Oporteat A + B cubo, cubum ex A == B demere. Abs ſolidis ſingularibus, quibus conſtat componendus abs A + B cubus, demantur ſingularia ſolida, quibus conſtat cubus abs A == B: orietur A quadratum in B ſexies, + B cubo bis. Hinc

THEOREMA.

CUBUS adgregati duorum laterum minus cubo differentiæ eorundem, æquatur ſextuplo ſolido à latere minore in quadratum majoris, plus duplo cubo lateris minoris.

PROPOSITIO XVII.

DIFFERENTIAM duorum laterum in tria ſingularia plana, quibus conſtat quadratum adgregati ipſorum laterum ſemel ſumpta, ducere.

SIT latus majus A, minus B. Oporteat A — B ducere in A quadratum, + A in B, + B quadrato. Fiat particularis ductio, & colligantur ſingularia ſolida. Erunt illa A cubus, — B cubo. Hinc

THEOREMA.

QUOD fit ex differentia duorum laterum in tria ſingularia plana, quibus conſtat quadratum adgregati ipſorum laterum ſemel ſumpta, æquale eſt differentiæ cuborum.

CONSECTARIUM.

DIFFERENTIA cuborum ſi adplicetur ad differentiam laterum, orientur tria ſingularia plana, quibus conſtat quadratum adgregati laterum ſemel ſumpta. Et permutim,

DIFFERENTIA cuborum ſi adplicetur ad tria ſingularia plana, quibus conſtat quadratum adgregati laterum ſemel ſumpta, orietur differentia laterum.

PROPOSITIO XVIII.

ADGREGATUM duorum laterum in tria ſingularia plana, quibus conſtat quadratum differentiæ ipſorum laterum ſemel ſumpta, ducere.

SIT

Figure 2. (Vietae 1646) p. 20

[5] "Itaque Proportio potest dici constitution æqualitatis. Æqualitas, resolutio proportionis" (Vietae 1646) p. 2.

This formula is essential to the scientific status of Viète's art because it connects his analytic techniques with the theorem-proving side of mathematics. As one would expect given this fundamental principle, he argues in succeeding chapters for algebraic results on grounds drawn from the Euclidean theory of proportion. Like Cardano, then, Viète presented himself as a follower of Euclid. However, he appeals to book V only, and since this concerns proportion rather than geometry, it does not impose the same constraints as Cardano's standards of rigour. It is, therefore, less misleading to describe him as a follower of Eudoxus than as a follower of Euclid.

In Viète, then, we have on the one hand an art with an aspiration to scientific status grounded by a sort of equivalence principle in Euclidean (or better, Eudoxan) tradition and presented as a body of theorems. On the other, we have a useful notational innovation that permits one to argue for general results, but which has not yet developed to the point where the manipulation of symbolic expressions counts as argument.

3. THOMAS HARRIOT (1560-1618)

Thomas Harriot's *Artis Analyticae Praxis* (1631) shares the scientific aspirations of Viète's algebra. Like Viète, Harriot lays down definitions and he orders his results into minor lemmas and major theorems. He celebrates a completed proof with *Quad Erat Probandum* or *Quad Erat Demonstrandum*. In spite of the title given to his work when it was published posthumously (*Artis*...), he evidently shared Viète's aspiration to establish algebra as a science. There is another similarity to Viète that connects Harriot with book V of Euclid. Look at his proof that the arithmetic mean of two unequal numbers is greater than the geometric mean (figure 3). Rather than start with the minimal assumption that $p>q$, and then multiply each side by $p-q$, he observes that p^2, pq, q^2 is a series in proportion and therefore he can immediately state that $p^2-pq > pq-q^2$. As in Viète, the theory of proportions in book V of Euclid is the assumed ground upon which the new science is to be built.

Another aspect of Harriot's scientific aspirations for algebra was his system of canonical equations. These are equations with straightforward solutions. Every non-canonical equation was associated with a canonical equation of the same degree, from which, Harriot hoped, it would be possible to solve the non-canonical case, or at least calculate the number and signs of the solutions. Harriot was not the first to attempt to classify equations. The point is that an objective taxonomy is part of what would distinguish a true science from an art or craft.

Harriot's notation was more highly developed than that of Viète. He employed Robert Recorde's familiar equality sign and the present-day symbol for inequality. He did not use superscripted numbers for exponents (we owe that to Descartes[6]), but instead repeated the letter or expression: for our p^2 he wrote *pp*. Consequently, he could not have entertained fractional or complex exponents. Rather than use brackets, he expressed products of polynomials by listing the factors vertically, next to a vertical line.

That is, for our *(a+b)(c-d)*, Harriot wrote $\left.\begin{matrix} a+b \\ c-d \end{matrix}\right|$

(it is easy to imagine how this might have evolved out of the usual method for multiplying numbers). This capacity of Harriot's notation to combine symbols for operations made it sufficiently powerful to induce a qualitative difference from the work of Cardano and Viète. In Cardano, mathematical argument, expressed in Latin prose, made essential use of diagrams and appealed to geometrical intuition. As for Viète, what little argument he offers is couched in a mixture of symbols and prose. That is because he had nothing to play the role of brackets. His notation did not allow him to combine operations and thereby create complex expressions. Rather, to avoid ambiguity, he had to fall back into prose. To return to the example above, $6A^2B+2B^3$ is written as "A quadratum in B sexies + B cubo bis". However, replacing a complex expression with an equivalent is the characteristic form of algebraic argument. It is precisely because Viète's notation did not allow this that he was unable to lay out his derivations explicitly, relying instead on his readers' numerical intuition. In Harriot, however, *the manipulation of symbols counts as argument*. The argument starts with the condition of the theorem expressed as an equation (or in the case in hand, an inequality). By a series of truth-preserving manipulations this is converted into the required conclusion. Cardano's great slabs of Latin prose have vanished, replaced by a terse commentary, the sole purpose of

6 Descartes does not appear in this photo-album for two reasons. First, in his mathematics the relation between algebra and geometry is too complex and philosophically fraught to be treated briefly. Second, in his epistemology he rejects formal deduction in favour of the lucid perception of clear and distinct ideas. For him, a deductive sequence is one intellectual intuition sliced into a series of lesser intuitions in order to accommodate the limits of human memory. Consequently, to divide the movement from premises to conclusion into a sequence of small steps can have pedagogic value only. By his own standards he had no logical requirement to do this: readers in whom the light of reason burns sufficiently brightly will see the truth of his claims. For more on Descartes and deduction see (Larvor 2001)

which is to distinguish each manipulation from its neighbours ("*Ergo*... *Sed*...").

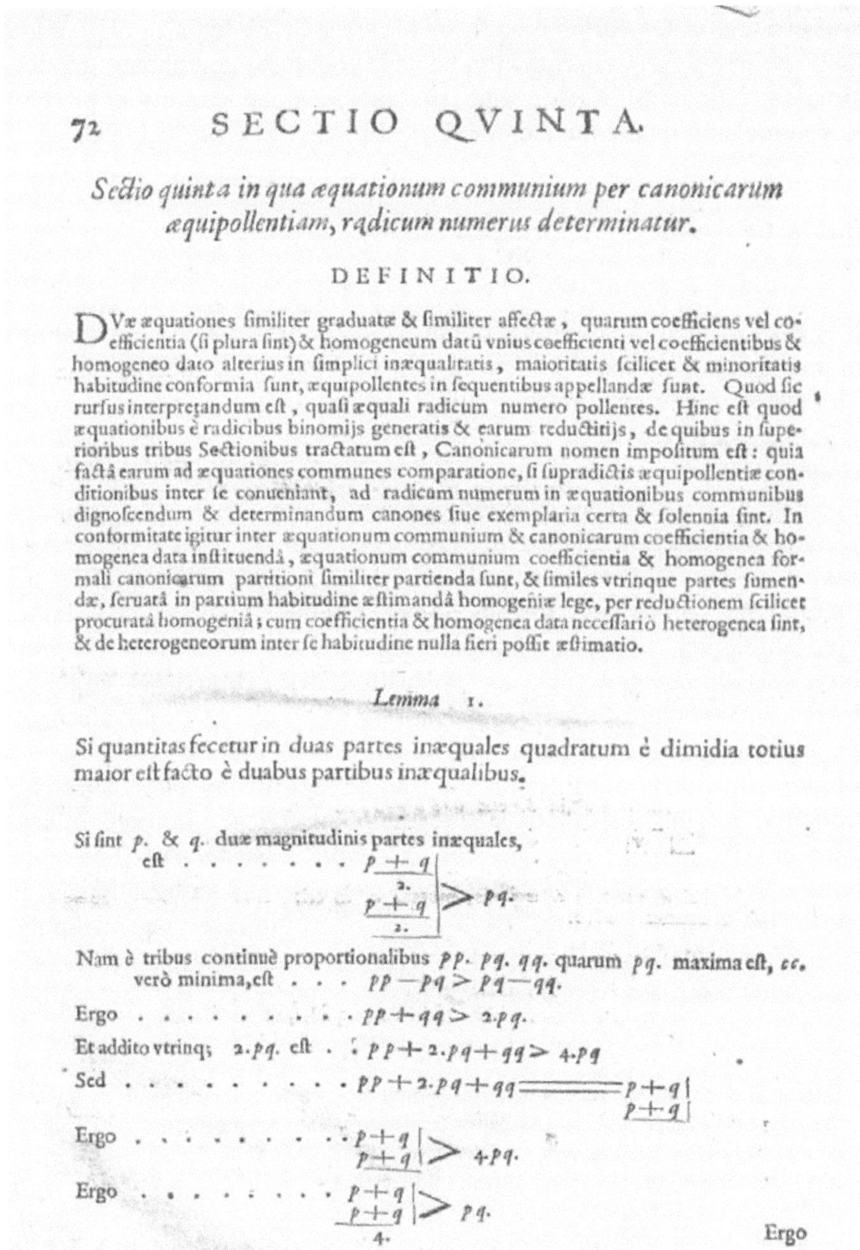

72 SECTIO QVINTA.

Sectio quinta in qua æquationum communium per canonicarum æquipollentiam, radicum numerus determinatur.

DEFINITIO.

DVæ æquationes ſimiliter graduatæ & ſimiliter affectæ, quarum coefficiens vel coefficientia (ſi plura ſint) & homogeneum datũ vnius coefficienti vel coefficientibus & homogeneo dato alterius in ſimplici inæqualitatis, maioritatis ſcilicet & minoritatis habitudine conformia ſunt, æquipollentes in ſequentibus appellandæ ſunt. Quod ſic rurſus interpretandum eſt, quaſi æquali radicum numero pollentes. Hinc eſt quod æquationibus è radicibus binomijs generatis & earum reductitijs, de quibus in ſuperioribus tribus Sectionibus tractatum eſt, Canonicarum nomen impoſitum eſt: quia factâ earum ad æquationes communes comparatione, ſi ſupradictis æquipollentiæ conditionibus inter ſe conueniant, ad radicum numerum in æquationibus communibus dignoſcendum & determinandum canones ſiue exemplaria certa & ſolennia ſint. In conformitate igitur inter æquationum communium & canonicarum coefficientia & homogenea data inſtituendâ, æquationum communium coefficientia & homogenea formali canonicarum partitioni ſimiliter partienda ſunt, & ſimiles vtrinque partes ſumendæ, ſeruatâ in partium habitudine æſtimandâ homogeniæ lege, per reductionem ſcilicet procuratâ homogeniâ; cum coefficientia & homogenea data neceſſariò heterogenea ſint, & de heterogeneorum inter ſe habitudine nulla fieri poſſit æſtimatio.

Lemma 1.

Si quantitas ſecetur in duas partes inæquales quadratum è dimidia totius maior eſt facto è duabus partibus inæqualibus.

Si ſint $p.$ & $q.$ duæ magnitudinis partes inæquales,
eſt . . . . $\frac{p+q}{2.} \cdot \frac{p+q}{2.} > pq.$

Nam è tribus continuè proportionalibus $pp.\ pq.\ qq.$ quarum $pq.$ maxima eſt, $cc.$ verò minima, eſt . . . $pp - pq > pq - qq.$

Ergo . . . . $pp + qq > 2.pq.$

Et addito vtrinq; $2.pq.$ eſt . . $pp + 2.pq + qq > 4.pq$

Sed . . . . $pp + 2.pq + qq = (p+q)(p+q)$

Ergo . . . . $(p+q)(p+q) > 4.pq.$

Ergo . . . . $\frac{(p+q)(p+q)}{4.} > pq.$

Ergo

Figure 3. (Harriot 1631) p. 72

It is at about this point that mathematics became ‘the discipline in which your pencil is smarter than you are’. That is, insight is still required but where this fails, one can go some distance by relying on the notation and its rules (though these are as yet implicit).

4. JOHN PELL (1611-1685)

The final photograph is of the work of John Pell. This is inevitably a blurred image because Pell did not publish a work on algebra under his own name. Instead, we have *An Introduction to Algebra* (1668), originally written in German by a Johann Rahn, translated into English by one Thomas Brancker “much altered and augmented by D. *P*.” (title page). Whatever the division of authorship between Rahn, Brancker and Pell, this work is interesting because in it the passage of algebra from geometry and prose to symbolic manipulation reaches its conclusion (though the development of the notation does not).

On the page reproduced here as figure 4, we find the recipe for Pythagorean triads. This is just one of many number-theoretic results in this work. The interest for our purposes is the division of the page into three columns. On the right, in the broadest column, the proof is carried out. In the narrow, middle column, each line of the proof has a number. In the left-hand column, each line has a note using previous line-numbers to explain how the current line of the proof was derived. These elements—numbered lines and comments—are now familiar to programmers and logicians. The comments themselves are easily standardised because there are only finitely many possible types of manipulation to get from one line to the next, namely the usual operations of arithmetic, plus the substitution of equivalent expressions. In place of Cardano’s geometric standards of rigour, we now have a simple rule: make sure you apply the same operation to both sides of the equation. One could, in principle, check this syntactically.

This possibility was not lost on one of the leading mathematicians of the day. Leibniz dreamed of a language in which logical errors would show up as faulty grammar, as if in natural language “errors were due to solecisms or barbarisms”.[7] The development of algebraic notation meant that algebra had

7 *De scientia universalis seu calculo philosophico* (Leibniz and Gerhardt 1960) volume VII, page 200 : “Id scilicet efficiendum est, ut omnis paralogismus nihil aliud sit quam error calculi, et ud sophisma, in hoc novae scripturae genere expressum, revera nihil aliud sit quam soloecismus vel barbarismus, ex ipsis grammatices hujus philosophicae legibus facile revincendus.”

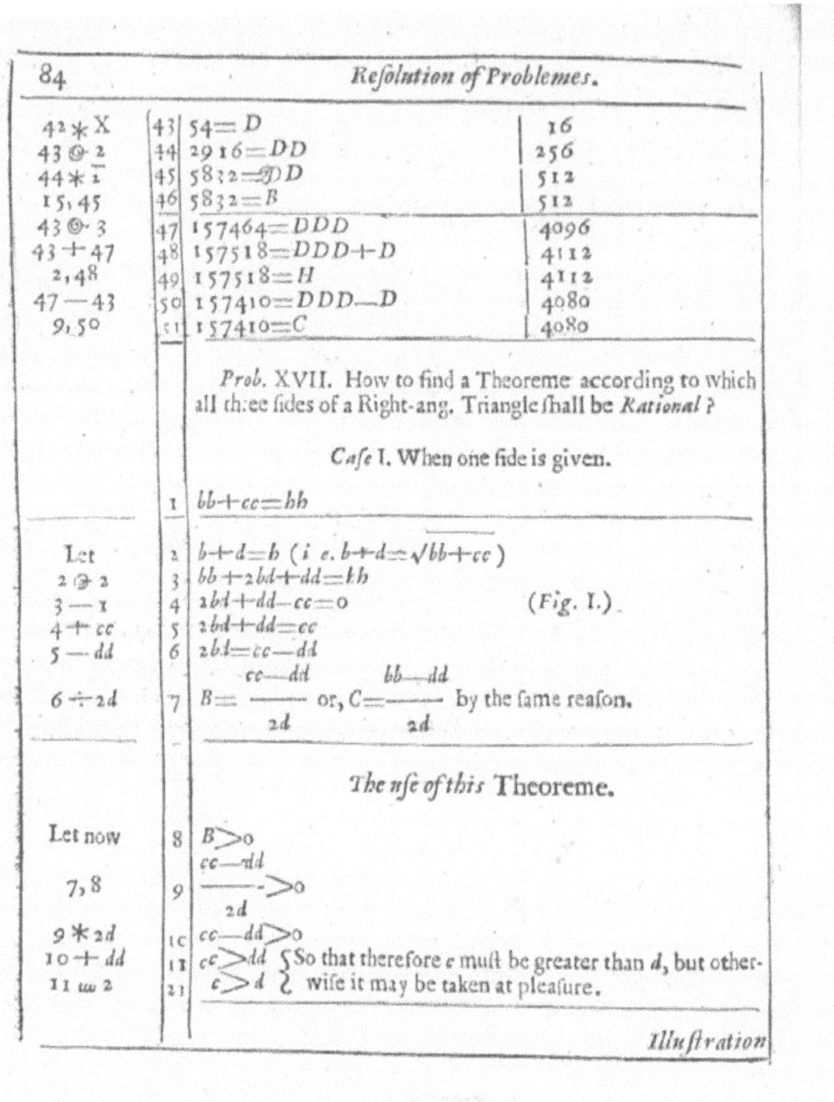

84 *Resolution of Problemes.*

42 ✱ X	43	54=D	16
43 ⊙ 2	44	2916=DD	256
44 ✱ $\overline{1}$	45	5832=DD	512
15, 45	46	5832=B	512
43 ⊙ 3	47	157464=DDD	4096
43+47	48	157518=DDD+D	4112
2, 48	49	157518=H	4112
47−43	50	157410=DDD−D	4080
9, 50	51	157410=C	4080

Prob. XVII. How to find a Theoreme according to which all three ſides of a Right-ang. Triangle ſhall be *Rational* ?

Caſe I. When one ſide is given.

	1	$bb+cc=hh$
Let	2	$b+d=h$ (*i e.* $b+d=\sqrt{bb+cc}$)
2 ⊙ 2	3	$bb+2bd+dd=hh$
3 − 1	4	$2bd+dd-cc=0$ (*Fig.* I.)
4 + cc	5	$2bd+dd=cc$
5 − dd	6	$2bd=cc-dd$
6 ÷ 2d	7	$B=\frac{cc-dd}{2d}$ or, $C=\frac{bb-dd}{2d}$ by the ſame reaſon.

The uſe of this Theoreme.

Let now	8	$B>0$
7, 8	9	$\frac{cc-dd}{2d}>0$
9 ✱ 2d	10	$cc-dd>0$
10 + dd	11	$cc>dd$ So that therefore *c* muſt be greater than *d*, but otherwiſe it may be taken at pleaſure.
11 ω 2	12	$c>d$

Illuſtration

Figure 4. (Rahn Johann, Brancker et al. 1668) p. 84

this property (or something very close to it) by the middle of the seventeenth century.

That is, Cardano could have expressed faulty mathematical arguments in correct and elegant Latin, because there was no relationship between the syntax of the language and the rigour of the proof. Indeed, the science-minded philosophers of the seventeenth century tended to suspect that the literary skill of their humanist predecessors served precisely to disguise logical fallacies. In the language of algebra, however, faulty logic shows up as faulty syntax. It would be a long time before anyone was in a position to try to prove this, not least because the modern notion of a wholly uninterpreted symbolism was not yet fully articulated. Nevertheless, the possibility of automated ('blind'), valid argument was discernible (to Leibniz at any rate) in the algebra of the mid-to-late seventeenth century.

5. DISPUTES OVER THE NOTION OF PROOF

Our sequence of four snapshots illustrates a change in the notion of mathematical proof in little over a century between Cardano (1545) and Pell (1668). Indeed, the crucial change was already present in Harriot's posthumous work of 1631, in which the manipulation of symbols is presented as proof. As we have seen, such manipulation was possible because Harriot's notation had a device equivalent to brackets, so he could distribute multiplication over addition. This richness permitted the substitution of equivalent expressions. Gathering terms and multiplying out brackets could now be done explicitly in the notation rather than merely described in prose. Therefore, the introduction of brackets is as important a step as Viète's use of letters for unknowns and coefficients. This shift in the notion of proof required the abandonment of Cardano's conception of rigour, dependant as it was on geometrical intuition. This change was part of the much larger philosophical and scientific turmoil of the time—Cardano published his *Ars Magna* two years after the publication of Copernicus' *De revolutionibus orbium coelestium*. In view of the intense contemporary debate about the philosophical foundations of scientific knowledge, it is implausible that mathematicians of the time did not ask whether these changes were compatible with rigour, even though our four chosen figures had relatively little to say about the nature of proof. Mention has already been made of the renaissance *Quaestio de Certitudine Mathematicarum*. As the seventeenth century opened the debate shifted its focus. Rather than accounting for the certainty of (Euclidean) mathematics, the problem was to understand the logical relations between different parts of mathematics. Algebra was only one new arrival: mathematicians had to contemplate complex numbers and indivisibles too. These changes have generated a rich historiography[8], to which I hope to add just one point.

[8] See (Mancosu 1992; Mancosu 1996) for developments in France and Italy; (Sasaki 1985; Pycior 1997) for the English end of the story, and the Hobbes-Wallis-Barrow controversy in particular.

There was, by the middle of the seventeenth century, an accumulation of results that could not possibly be proved geometrically. These were mostly number-theoretic results such as the sums of finite series (principally arithmetic progressions and powers thereof), or formulae such as the Pythagorean recipe taken from Pell's algebra above. Mathematicians explored the triangle of binomial coefficients ('Pascal's triangle'). Some of these (such as the Pythagorean recipe) could be proved using existing algebraic techniques, while others would have to wait for the development of proof by complete induction. In addition, equations themselves were becoming objects of study. Cardano noted that the complex roots of an equation occur in pairs, but had no means to prove it. The symmetric functions of coefficients were identified in Girard's *L'invention nouvelle en l'algèbre* of 1629.[9] Descartes' 'rule of signs' for calculating the number of positive and negative roots of a real-valued equation from the changes in sign in the coefficients (intimated in Cardano and sometimes attributed to Harriot) had to wait until the eighteenth century for a rigorous treatment. In his algebra, Pell appealed to the modern sense of dimension: the number of data should equal the number of unknowns, or else the solution will be under- or over-determined.[10] Cardano had to admit that a few rather recondite phenomena could not be treated within his standards of rigour. Since these were so few, he had the option of conserving his standards while noting the anomalies as such. His successors in the following century faced an avalanche of arithmetical and number theoretical results that could not be proved in Cardano's Euclidean style.

Moreover, algebra began to distinguish itself from arithmetic as mathematicians discerned general features of equations (such as the symmetric functions or the rule of signs). As a result, the conservative horn of Cardano's dilemma disappeared. Faced with all this new material, mathematicians had no option than to abandon Cardano's geometrical standards of rigour.

REFERENCES

Cardano, G. (1570). *Opvs novvm de proportionibvs nvmerorvm, motvvm, pondervm, sonorvm, aliarvmqve rervm mensurandarum, non solùm geometrico more stabilitum, sed etiam uarijs experimentis & obseruationibus rerum in natura ... Praeterea Artis magnae; sive, De regvlis algebraicis, liber vnvs, abstrvsissimvs & inexhaustus plane totius arithmeticae thesaurus, ab authore recens multis in locis recognitus & auctus. Item, De aliza regvla liber, hoc est, algebraicae logisticae suae, numeros recondita numerandi subtilitate,*

[9] (Girard and Haan 1884).

[10] (Wallis 1685), p. 214.

secundum geometricas quantitates inquirentis, necessaria coronis, nunc demum in lucem edita. Basileae, Ex officina Henricpetrina.

Girard, A. and D. B. d. Haan (1884). *Invention nouvelle en l'algáebre.* Leiden,, Imprimâe chez Murâe fráeres.

Harriot, T. (1631). *Artis analyticae praxis, ad Aequationes algebraicas nova, expedita & generali methodo resolvendas.* London, Robert Barker.

Larvor, B. (2001). "Old Maps, Crystal Spheres and the Cartesian Circle." *Graduate Faculty Philosophy Journal* **22**(2).

Leibniz, G. W. and K. I. Gerhardt (1960). *Die philosophischen Schriften von Gottfried Wilhelm Leibniz.* Hrsg. von C.I. Gerhardt. Hildesheim, G. Olms Verlagsbuchhandlung.

Mancosu, P. (1992). "Aristotelian Logic and Euclidean Mathematics: Seventeenth-century Developments of the *Quaestio de Certitudine Mathematicarum.*" *Studies in the History and Philosophy of Science* **23**(2): 241-265.

Mancosu, P. (1996). *Philosophy of Mathematics and Mathematical Practice in the Seventeenth Century.* New York; Oxford, Oxford University Press.

Pycior, H. M. (1997). *Symbols, impossible numbers, and geometric entanglements : British algebra through the commentaries on Newton's Universal arithmetick.* Cambridge, U.K ; New York, N.Y., Cambridge University Press.

Rahn Johann, H., T. Brancker, et al. (1668). *An introduction to Algebra, translated out of the High-Dutch into English, by T. Branckner ... Much altered and augmented by D. P. [i.e. Dr. John Pell.] Also a table of Odd Numbers less than one hundred thousand, shewing those that are incomposit, and resolving the rest into their factors or coefficients, &c.* London, W. G. for Moses Pitt.

Sasaki, C. (1985). "The Acceptance of the Theory of Proportion in the Sixteenth and Seventeenth Centuries--Barrow's reaction to the Analytic Mathematics." *Historia Scientiarum* **29**: 83-116.

Vietae, F. (1646). *Opera mathematica in unam volumen congesta ac recognita.* Lugdunum Batavorum,, Elzevir.

Viete, F. o. and T. R. Witmer (1983). *The analytic art: nine studies in algebra, geometry and trigonometry from the Opus restitutae mathematicae analyseos, seu algebra nova.* Kent (Ohio), Kent State University Press.

Wallis, John (1685). *A treatise of algebra, both historical and practical.* London: John Playford.

Witmer, T. R. and G. Cardano (1968). *The great art: or, The rules of algebra.* Cambridge, Massachusetts; London, the M.I.T. Press.

Chapter 8

THE INFORMAL LOGIC OF MATHEMATICAL PROOF

Andrew Aberdein
Florida Institute of Technology

Abstract: Informal and formal logic are complementary methods of argument analysis. Informal logic provides a pragmatic treatment of features of argumentation which cannot be reduced to logical form. This paper shows how paying attention to aspects of mathematical argumentation captured by informal, but not formal, logic can offer a more nuanced understanding of mathematical proof and discovery.

Key words: dialectic, four colour theorem, informal logic, mathematical proof, Stephen Toulmin, Douglas Walton

The proof of mathematical theorems is central to mathematical practice and to much recent debate about the nature of mathematics: as Paul Erdös often remarked, 'a mathematician is a machine for turning coffee into theorems' [9, p. 7]. This paper is an attempt to introduce a new perspective on the argumentation characteristic of mathematical proof. I shall argue that this account, an application of informal logic to mathematics, helps to clarify and resolve several important philosophical difficulties.

It might be objected that formal, deductive logic tells us everything we need to know about mathematical argumentation. I shall leave it to others [14, for example] to address this concern in detail. However, even the protagonists of explicit reductionist programmes—such as logicists in the philosophy of mathematics and the formal theorem proving community in computer science—would readily concede that their work is not an attempt to capture actual mathematical practice. Having said that, mathematical argumentation is certainly not inductive either. Mathematical proofs do not involve inference from particular observations to general laws. A satisfactory account of mathematical argumentation must include deductive inference, even if it is not exhausted by it. It must be complementary, rather

B. Van Kerkhove and J.P. van Bendegem (eds.), Perspectives on Mathematical Practices, 135–151.

than hostile, to formal logic. My contention is that a suitable candidate has already been developed independently: informal logic.

Informal logic is concerned with all aspects of inference, including those which cannot be captured by logical form. It is an ancient subject, but has been a degenerating research programme for a long time. Since the nineteenth century it has been overshadowed by the growth of formal logic. More fundamentally, it has suffered by identification with the simplistic enumeration of fallacies, without any indication of the circumstances in which they are illegitimate. Since most fallacies can be exemplified in some contexts by persuasive, indeed valid, arguments, this approach is of limited use. In recent decades more interesting theories have been developed. I shall look at two of the most influential, and discuss their usefulness for the analysis of mathematical proof.

1. TOULMIN'S PATTERN OF ARGUMENT

One of the first modern accounts of argumentation is that developed in Stephen Toulmin's *The Uses of Argument* [18]. Toulmin offers a general account of the layout of an argument, as a claim (C) derived from data (D), in respect of a warrant (W). Warrants are general hypothetical statements of the form 'Given D, one may take it that C' [18, p. 99]. Hence the laws of logic provide a warrant for deductive inferences. However, the pattern is intended to be more general, and provides for different, weaker warrants, although these would not permit us to ascribe the same degree of certainty to C. This is recognized by the inclusion of a modal qualifier (Q), such as

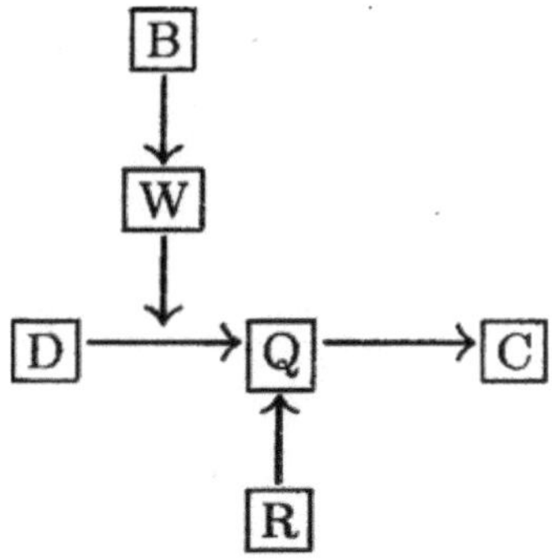

Figure 1. Toulmin's DWC pattern

'necessarily', 'probably', 'presumably', …, in the pattern. If the warrant is defeasible, we may also specify the conditions (R) under which it may be rebutted. Finally, the argument may turn on the backing (B) which can be provided for W. Toulmin's claim is that the general structure of a disparate variety of arguments may be represented as in Figure 1.

Interpreting the letters as above, this diagram may be read as follows: "Given D, we can (modulo Q) claim C, since W (on account of B), unless R". In Toulmin's vintage example: "Given that Harry was born in Bermuda, we can presumably claim that he is British, since anyone born in Bermuda will generally be British (on account of various statutes ...), unless he's a naturalized American, or his parents were aliens, or ...". In simpler examples B, Q and R may not all be present, but D, W and C are taken to be essential to any argument, hence the description of this model as the DWC pattern. Toulmin stresses the field dependency of the canons of good argument: what counts as convincing may vary substantially between the law court, the laboratory and the debating chamber. In particular, what counts as acceptable backing will turn significantly on the field in which the argument is conducted [18, p. 104]. Toulmin's work has been very influential in the study of argument, despite an initially chilly reception amongst philosophers and logicians.[1] It was quickly adopted by communication theorists, after the publication in 1960 of a celebrated paper by Wayne Brockriede and Douglas Ehninger [4]. Toulmin's account of argumentation is now the dominant model in this field. More recently, his work has been widely studied by computer scientists attempting to model natural argumentation [13, for example]. One purpose of Toulmin's critique of logic is to dispute the utility of formal logic for the analysis of any argumentational discourse *other* than mathematics: he regards mathematical proof as one of the few success stories for the formal logic tradition. Nevertheless, mathematical proofs can be subsumed under the DWC pattern, where the warrant is backed by various axioms, rules of inference and mathematical techniques providing grounds for supposing the claim to be necessary, given the data. Toulmin provides an example in a later collaborative work [19, p. 89], by reconstructing Theaetetus's proof that there are exactly five polyhedra. The data and warrant consist of various facts about the platonic solids, the warrant is backed by the axioms, postulates and definitions of three-dimensional Euclidean geometry, and the modal qualifier 'with strict geometrical necessity' admits of no rebuttal or exception within the bounds of Euclidean geometry.

1 'Unanimous … condemnation' according to van Eemeren *et al.* [7, p. 164], who survey the book's reviewers.

2. APPLYING TOULMIN TO MATHEMATICS

The significance of Toulmin's work for mathematical proof is explored at greater length in what I believe to be the only study so far of the application of informal logic to mathematics, a paper written in Catalan by Jesus Alcolea Banegas [1][2] Alcolea makes use of a further distinction of Toulmin's, introduced in [19]: that between *regular* and *critical* arguments. This distinction echoes Thomas Kuhn's contrast between normal and revolutionary science: a regular argument is an argument within a field which appeals to the already well-established warrants characteristic of the field, whereas a critical argument is an argument used to challenge prevailing ideas, focusing attention on the assumptions which provide a backing for the warrants of regular arguments. Critical arguments must therefore appeal to different warrants. Mathematical proofs are regular arguments, although they may give rise to critical arguments if they are especially interesting or controversial. Conversely, metamathematical debates are critical arguments, but they often provide new opportunities for proofs, that is, regular arguments.

Alcolea uses Toulmin's layout to reconstruct one regular and one critical argument from mathematics. The critical argument, the debate over the admissibility of the axiom of choice, is the more fully developed and persuasive of Alcolea's case studies. It is perhaps not too surprising that critical arguments in mathematics are similar to critical arguments in the other sciences, since ultimately they are not arguments in mathematics, but arguments *about* mathematics, that is to say they are metamathematical. However, my concern is primarily with the argumentation of mathematics itself, rather than that of metamathematics. Hence I shall concentrate on Alcolea's example of a regular argument: Kenneth Appel and Wolfgang Haken's proof of the four colour conjecture.[3] He reconstructs the central argument of the proof as a derivation from the data D1–D3

(D1) Any planar map can be coloured with five colours.
(D2) There are some maps for which three colours are insufficient.
(D3) A computer has analysed every type of planar map and verified that each of them is 4-colourable.

of the claim C, that

2 I am grateful to Miguel Gimenez of the University of Edinburgh for translating this paper.
3 For further detail of the proof see [2], [25] or [12].

(C) Four colours suffice to colour any planar map.

by employment of the warrant W, which has backing B

(W) The computer has been properly programmed and its hardware has no defects.
(B) Technology and computer programming are sufficiently reliable. [1, pp. 142f.]

He regards this as making clear that, since the warrant is not wholly mathematical, the proof must leave open the possibility of 'a specific counterexample, that is to say, a particular map that cannot be coloured with four colours might still exist' [1, p. 143].[4]

This example demonstrates both the strengths and the dangers of this approach. To complete Toulmin's layout we are obliged to make explicit not merely the premises and the conclusion, but also the nature of the support which the former is supposed to lend the latter. Thus the focus of Appel and Haken's critics, the heterodox deployment of a computer in a mathematical proof, is made glaringly obvious. However, it is premature to draw from this surface dissimilarity the inference that Appel and Haken's result is less convincing than other mathematical proofs. A closer reading of Alcolea's reconstruction may clarify this point. Premises D1 and D2 have conventional mathematical proofs, as Alcolea points out. (D1 is not strictly relevant to the derivation of C, although its proof originated techniques which were instrumental to Appel and Haken's work.) D3 is a very concise summary of Appel and Haken's central results. It may help to spell out the details at greater length.

There are two essential ideas behind the Appel and Haken proof: unavoidability and reducibility. An unavoidable set is a set of configurations, that is countries or groups of adjacent countries, at least one of which must be present in any planar map. For example, all such maps must contain either a two-sided, a three-sided, a four-sided or a five-sided country, so these configurations constitute an unavoidable set. A configuration is reducible if any map containing it may be shown to be four-colourable. Two-sided, three-sided, and four-sided countries are all reducible. To prove the four colour theorem it suffices to exhibit an unavoidable set of reducible configurations. Alfred Kempe, who introduced the concepts of unavoidability and reducibility, was believed to have proved the four colour theorem in 1879 by showing that five-sided countries were also reducible

4 '. . . un contraexemple especific, és a dir, que es trobe un mapa particular que no puga colarar-se amb quatre colors'

[11]. However, in 1890 a flaw was discovered in his reasoning: the five-sided country is not reducible, hence a larger unavoidable set is required if all its configurations are to be reducible. Appel and Haken used a computer to search for such a set, eventually discovering one with 1,482 members. The unavoidability of this set could be demonstrated by hand, but confirming the reducibility of all its members would be far too protracted a task for human verification. Subsequent independent searches have turned up other unavoidable sets. The smallest to date is a set of 633 reducible configurations found by Neil Robertson, Daniel Sanders, Paul Seymour and Robin Thomas in 1994.[5] Verifying the reducibility of these configurations still requires a computer.

So for there to be a non-four-colourable planar map, as Alcolea suggests, Appel and Haken (and their successors) must have erred either in the identification of the unavoidable set, or in the demonstration of the reducibility of its member configurations. Since the former step can be verified by conventional methods, the computer can only be suspected of error in demonstrating reducibility. Two sorts of computer error should be distinguished: a mistake may be made in the programming, or a fault may arise in the computer itself (the hardware or firmware). The former error would arise due to a human failure to correctly represent the mathematical algorithms which the computer was programmed to implement. This sort of mistake does not seem to be interestingly different from the traditional type of mathematical mistake, such as that made by Kempe in his attempt to prove the four colour conjecture. The second sort of error is genuinely new. However, it would seem to be profoundly unlikely.

Computer hardware can exhibit persistent faults, some of which can be hard to detect.[6] However, the potential risks of such faults can be minimized by running the program on many different machines. One might still worry about Appel and Haken's programs, since they were written in machine code and would therefore be implemented in more or less exactly the same manner on any computer capable of running them, perhaps falling foul of the same bug each time. This sort of checking might be suspected of being no better than buying two copies of the same newspaper to check the veracity of its reporting.[7] However, the same reducibility results were achieved independently, using different programs, as part of the refereeing process for Appel and Haken's work. Moreover, the more recent programs of Robertson *& al.* were written in higher level languages, as are the programs employed in most other computer-assisted proofs. The existence of different compilers

[5] For details of their publications, see [25, p. 244].

[6] For example, the notorious Pentium FDIV bug.

[7] As Wittgenstein once remarked in a different context: [26, § 265].

and different computer platforms ensures that these programs can be implemented in many intrinsically different ways, reducing the likelihood of hardware or firmware induced error to the astronomical.[8]

Thus we may derive an alternative reconstruction of Appel and Haken's argument: "Given that (D4) the elements of the set U are reducible, we can (Q) almost certainly claim that (C) four colours suffice to colour any planar map, since (W) U is an unavoidable set (on account of (B) conventional mathematical techniques), unless (R) there has been an error in either (i) our mathematical reasoning, or (ii) the hardware or firmware of all the computers on which the algorithm establishing D4 has been run." If, in addition, we observe that (i) appears to be orders of magnitude more likely than (ii), then C would seem to be in much less doubt than it did in the light of Alcolea's reconstruction. The purpose of the preceding has been not so much to rescue the four colour conjecture from Alcolea's critique (although few if any graph theorists would accept that a counterexample is possible), but to show up the limitations of Toulmin's pattern as a descriptive technique. As other critics have pointed out, reconstructing an argument along Toulmin's lines 'forces us to rip propositions out of context' [24, p. 318]. The degree of abstraction necessary to use the diagram at all can make different, incompatible, reconstructions possible, leaving the suspicion that any such reconstruction may involve considerable (and unquantified) distortion.[9]

3. WALTON'S NEW DIALECTIC

There has been significant progress in informal logic since the publication of *The uses of argument*. One milestone was the publication of Charles Hamblin's *Fallacies* [8] in 1970. This demonstrated the inadequacies of much of traditional fallacy theory and, by way of remedy, proposed an influential dialectical model of argumentation. Further impetus has come from the recent work of communication theorists such as Frans van Eemeren and Rob Grootendorst [6]. One contemporary logician who

[8] Georges Gonthier of Microsoft Research recently announced the successful verification of a proof similar to that of Robertson *& al.* in the proof checking software Coq. This would eliminate the risk of human error, and reduce the risk of computer error even further. See Georges Gonthier, 2005, 'A computer-checked proof of the Four Colour Theorem', http://research.microsoft.com/%7Egonthier/4colproof.pdf

[9] For an attempt to defend Toulmin from these charges, see my 'The uses of argument in mathematics', *Argumentation*, forthcoming.

shows the influence of both traditions is Douglas Walton.[10] The focus of his work is the dialectical context of argument. Walton distinguishes between 'inference', defined as a set of propositions, one of which is warranted by the others, 'reasoning', defined as a chain of inferences, and 'argument', defined as a dialogue employing reasoning. This dialectical component entails that arguments require more than one arguer: at the very least there must be an assumed audience, capable in principle of answering back. Winston Churchill once praised the argumentational skills of the celebrated barrister and politician F. E. Smith, 1st Earl of Birkenhead, by stressing their suitability to context: 'The bludgeon for the platform; the rapier for a personal dispute; the entangling net and unexpected trident for the Courts of Law; and a jug of clear spring water for an anxious perplexed conclave' [5, p. 176]. Toulmin also stresses the domain specificity of good practice in argument. What is distinctive about Walton's analysis is the attempt to characterize dialectical context in terms of general features which are not themselves domain specific. Without pretending to have an exhaustive classification of argumentational dialogue, he is able to use these features to draw several important distinctions. The principal features with which he is concerned are the 'initial situation' and the 'main goal' of the dialogue. The initial situation describes the circumstances which give rise to the dialogue, in particular the differing commitments of the interlocutors. The main goal is the collective outcome sought by both (all) participants, which may be distinct from their individual goals.

	Initial Situation		
Main Goal	Conflict	Open Problem	Unsatisfactory Spread of Information
Stable Agreement/ Resolution	Persuasion	Inquiry	Information Seeking
Practical Settlement/ Decision (Not) to Act	Negotiation	Deliberation	
Reaching a (Provisional) Accommodation	Eristic		

Table 1. Walton & Krabbe's 'Systematic survey of dialogue types' [23, p. 80]

10 Walton has published a great number of works on informal logic. [22] provides an overview of the general method common to many of them.

If we simplify the situation by permitting each discussant to regard some crucial proposition as either true, false or unknown, four possibilities emerge. Either (0) the discussants agree that the proposition is true (or that it is false), in which case there is no dispute; or (1) one of them takes it to be true and the other false, in which case they will be in direct conflict with each other; or (2) they both regard it as unknown, which may result in a dialogue as they attempt to find out whether it is true or false; or (3) one of them believes the proposition to be true (or false) but the other does not know which it is. Thus we may distinguish three types of initial situation from which an argumentational dialogue may arise: a conflict, an open problem, or an unsatisfactory spread of information. A conflict may produce

Type of Dialogue	**Initial Situation**	**Individual Goals of Participants**	**Collective Goal of Dialogue**	**Benefits**
Persuasion	Difference of opinion	Persuade other party	Resolve difference of opinion	Understand positions
Inquiry	Ignorance	Contribute findings	Prove or disprove conjecture	Obtain knowledge
Deliberation	Contemplation of future consequences	Promote personal goals	Act on a thoughtful basis	Formulate personal priorities
Negotiation	Conflict of interest	Maximize gains (self-interest)	Settlement (without undue inequity)	Harmony
Information-Seeking	One party lacks information	Obtain information	Transfer of knowledge	Help in goal activity
Quarrel (Eristic)	Personal conflict	Verbally hit out at and humiliate opponent	Reveal deeper conflict	Vent emotions
Debate	Adversarial	Persuade third party	Air strongest arguments for both sides	Spread information
Pedagogical	Ignorance of one party	Teaching and learning	Transfer of knowledge	Reserve transfer

Table 2. Walton's types of dialogue [21, p. 605]

several different types of dialogue depending on how complete a resolution is sought. For a stable outcome one interlocutor must persuade the other, but, even if such persuasion is impossible they may still seek to negotiate a practical compromise on which future action could be based. Or they may aim merely to clear the air by expressing their contrasting opinions, without hoping to do more than merely agree to disagree: a quarrel. These three goals—stable resolution, practical settlement and provisional accommodation—can also be applied to the other two initial situations, although not all three will be exemplified in each case. So open problems can lead to stable resolutions, or if this is not achievable, to practical settlement. However, provisional accommodation should not be necessary if the problem is genuinely open, since neither discussant will be committed to any specific view. Where the dialogue arises merely from the ignorance of one party then a stable resolution should always be achievable, obviating the other goals.

The interplay of these different types of initial situation and main goal thus allows Walton to identify six principal types of dialogue, Persuasion, Negotiation, Eristic, Inquiry, Deliberation and Information Seeking, which may be represented diagrammatically as in Table 1. The contrasting properties of these different types of dialogue are set out in Table 2. This table also states the individual goals of the interlocutors typical to each type, and includes two derivative types: the debate, a mixture of persuasion and eristic dialogue, and the pedagogical dialogue, a subtype of the information seeking dialogue. Many other familiar argumentational contexts may be represented in terms of Walton's six basic types of dialogue by such hybridization and subdivision.[11]

It is central to Walton's work that the legitimacy of an argument should be assessed in the context of its use: what is appropriate in a quarrel may be inappropriate in an inquiry, and so forth. Although some forms of argument are never legitimate (or never illegitimate), most are appropriate if and only if they are "in the right place". For example, threats are inappropriate as a form of persuasion, but they can be essential in negotiation. In an impressive sequence of books, Walton has analyzed a wide variety of fallacious or otherwise illicit argumentation as the deployment of strategies which are sometimes admissible in contexts in which they are inadmissible. However, Walton has not directly addressed mathematical argumentation. In the next section I shall set out to explore how well his system may be adapted to this purpose.

4. APPLYING WALTON TO MATHEMATICS

In what context (or contexts) do mathematical proofs occur? The obvious answer is that mathematical proof is a special case of inquiry. Indeed, Walton states that the collective goal of inquiry is to 'prove or disprove [a] conjecture'. An inquiry dialogue proceeds from an open problem to a stable agreement. That is to say from an initial situation of mutual ignorance, or at

[11] See Table 3.1 in [23, p. 66] for some further examples.

least lack of commitment for or against the proposition at issue, to a main goal of shared endorsement or rejection of the proposition. This reflects a standard way of reading mathematical proofs: the prover begins from a position of open-mindedness towards the conjecture, shared with his audience. He then derives the conjecture from results upon which they both agree, by methods which they both accept.

But this is not the only sort of dialogue in which a mathematical proof may be set out. As William Thurston has remarked, mathematicians 'prove things in a certain context and address them to a certain audience' [17, p. 175]. Indeed, crucially, there are several different audiences for any mathematical proof, with different goals. Satisfying the goals of one audience need not satisfy those of the others. For example, a proof may be read by:

- Journal referees, who have a professional obligation to play devil's advocate;
- Professional mathematicians in the same field, who may be expected to quickly identify the new idea(s) that the proof contains, grasping them with only a few cues, but who may already have a strong commitment to the falsehood of the conjecture;
- Professional mathematicians in other (presumably neighbouring) fields, who will need more careful and protracted exposition;
- Students and prospective future researchers in the field, who could be put off by too technical an appearance, or by the impression that all the important results have been achieved;
- Posterity, or in more mercenary terms, funding bodies: proof priority can be instrumental in establishing cudos with both.

This list suggests that the initial situation of a proof dialogue cannot always be characterized as mutual open-mindedness. Firstly, in some cases, the relationship between the prover and his audience will be one of conflict. If the conjecture is a controversial one, its prover will have to convince those who are committed to an incompatible view. And if an article is refereed thoroughly, the referees will be obliged to adopt an adversarial attitude, irrespective of their private views.

Secondly, as the later items indicate, proofs have a pedagogic purpose. Thurston relates his contrasting experiences in two fields to which he made substantial contributions. As a young mathematician, he proved many results in foliation theory using powerful new methods. However, his proofs were of a highly technical nature and did little to explain to the audience how they too might exploit the new techniques. As a result, the field evacuated: other mathematicians were afraid that by the time they had mastered Thurston's methods he would have proved all the important results. In later work, on Haken manifolds, he adopted a different approach. By concentrating on proving results which provided an infrastructure for the field, in a fashion which allowed others to acquire his methods, he was able to develop a community of mathematicians who could pursue the field further than he could alone. The price for this altruism was that he could not take all the credit for the major results. Proofs which succeed in the context Thurston advocates proceed from an initial situation closer to Walton's 'unsatisfactory spread of information'. This implies that information seeking is another context in which mathematical proofs may be articulated. Of course, the information which is being sought is not merely the conjecture being proved, but also the methods used to prove it.

An unsatisfactory spread of information, unlike a conflict or an open problem, is an intrinsically asymmetrical situation. We have seen that proofs can arise in dialogues wherein the prover possesses information sought by his interlocutors. Might there be circumstances in which we should describe as a proof a dialogue in which the prover is the information seeker? This is the question considered by Thomas Tymoczko [20, p. 71] and Yehuda Rav [14], to somewhat different ends. Tymoczko considers a community of Martian mathematicians who have amongst their number an unparalleled mathematical genius, Simon. Simon proves many important results, but states others without proof. Such is his prestige, that "Simon says" becomes accepted as a form of proof amongst the Martians. Rav considers a fantastical machine, Pythiagora, capable of answering mathematical questions instantaneously and infallibly. Both thought experiments consider the admission of a dialogue with an inscrutable but far better informed interlocutor as a possible method of proof.

In both cases we are invited to reject this admission, although, interestingly, for different reasons. Rav sees his scenario as suggesting that proof cannot be purely epistemic: if it were then Pythiagora would give us all that we needed, but Rav suggests that we would continue to seek conventional proofs for their other explanatory merits. He concludes that Pythiagora could not give us proof. Tymoczko draws an analogy between his thought experiment and the use of computers in proofs such as that of the four colour theorem. He argues that there is no formal difference between claims backed by computer and claims backed by Simon: they are both appeals to authority. The difference is that the computer can be a warranted authority. Hence, on Tymoczko's admittedly controversial reading, computer assisted proof is an information seeking dialogue between the prover and the computer.

So far we have seen that the initial situation of a proof dialogue can vary from that of an inquiry. What of the main goal—must this be restricted to stable resolution? Some recent commentators have felt the need for a less rigorous form of mathematics, with a goal closer to Walton's practical settlement. Arthur Jaffe and Frank Quinn [10] introduced the much discussed, if confusingly named, concept of 'theoretical mathematics'. They envisage a division of labour, analogous to that between theoretical and experimental physics, between conjectural or speculative mathematics and rigorous mathematics. Where traditional, rigorous mathematicians have theorems and proofs, theoretical mathematicians make do with 'conjectures' and 'supporting arguments'. This echoes an earlier suggestion by Edward Swart [16] that we should refrain from accepting as theorems results which depend upon lengthy arguments, whether by hand or computer, of which we cannot yet be wholly certain. He suggests that 'these additional entities could be called agnograms, meaning theoremlike statements that we have verified as best we can but whose truth is not known with the kind of assurance that we attach to theorems and about which we must thus remain, to some extent, agnostic' [16, p. 705]. In both cases the hope is that further progress will make good the shortfall: neither Jaffe and Quinn's conjectures nor Swart's agnograms are intended as replacements for rigorously proved theorems.

More radical critics of the accepted standards of mathematical rigour suggest that practical settlement can be a goal of proof and not merely of lesser, analogous activities. For instance, Doron Zeilberger [27] envisages a future of semi-rigorous (and ultimately non-rigorous) mathematics in which the ready availability by computer of near certainty reduces the pursuit of absolute certainty to a low resource allocation priority. Hence he predicts that a mathematical abstract of the future could read "We show, in a certain precise sense, that the Goldbach conjecture is true with probability larger than 0.99999, and that its complete truth could be determined with a budget of \$10 billion" [27, p. 980]. 'Proofs' of this sort explicitly eschew stable resolution for practical settlement. Thus Zeilberger is arguing that proofs could take the form of deliberation or negotiation.

The last of Walton's dialogue types is the eristic dialogue, in which no settlement is sought, merely a provisional accommodation in which the commitments of the parties are made explicit. This cannot be any sort of proof, since no conclusion is arrived at. But it is not completely without interest. A familiar diplomatic euphemism for a quarrel is "a full and frank exchange of views", and such activity does have genuine merit. Similarly, even failed mathematical proofs can be of use, especially if they clarify previously imprecise concepts, as we saw with Kempe's attempted proof of the four colour conjecture [11]. This process has something in common, if not with a quarrel, at least with a debate, which we saw to be a related type of dialogue.

To take stock, we have seen that most of Walton's dialogue types are reflected to some degree in mathematical proof. Table 3, an adaptation of Table 2, sets out the difference between the various types of proof dialogue introduced.

Type of Dialogue	Initial Situation	Main Goal	Goal of Prover	Goal of Interlocutor
Proof as Inquiry	Open-mindedness	Prove or disprove conjecture	Contribute to outcome	Obtain knowledge
Proof as Persuasion	Difference of opinion	Resolve difference of opinion with rigour	Persuade interlocutor	Persuade prover
Proof as Information-Seeking (Pedagogical)	Interlocutor lacks information	Transfer of knowledge	Disseminate knowledge of results & methods	Obtain knowledge
'Proof' as Information-Seeking (*e.g.* Tymoczko)	Prover lacks information	Transfer of knowledge	Obtain information	Presumably inscrutable
'Proof' as Deliberation (*e.g.* Swart)	Open-mindedness	Reach a provisional conclusion	Contribute to outcome	Obtain warranted belief
'Proof' as Negotiation (*e.g.* Zeilberger)	Difference of opinion	Exchange resources for a provisional conclusion	Contribute to outcome	Maximize value of exchange
'Proof' as Eristic/ Debate	Irreconcilable difference of opinion	Reveal deeper conflict	Clarify position	Clarify position

Table 3. *Some types of proof dialogue*

5. PROOF DIALOGUES

In this last section I shall explore how the classification of proof dialogues may help to clarify many of the problems that have arisen in the philosophical debate over the nature of mathematical proof. We can see that proofs may occur in several distinct types of dialogue, even if we do not count the suspect cases (the entries for Table 3 where 'proof' is in scare quotes). An ideal proof will succeed within inquiry, persuasion and pedagogic proof dialogues. Suboptimal proofs may fail to achieve the goals of at least one of these dialogue types. In some cases, this may be an acceptable, perhaps inevitable, shortcoming; in others it would fatally compromise the argument's claim to be accepted as a proof.

As Thurston's experience with foliation theory demonstrated, not every proof succeeds pedagogically. Proofs in newly explored areas are often hard to follow, and there are some results which have notoriously resisted all attempts at clarification or simplification.[12] Yet, if these proofs succeed in inquiry and persuasion dialogues, we have no hesitation in accepting them. Conversely, there are some 'proofs' which have a heuristic usefulness in education, but which would not convince a more seasoned audience. Pedagogic success is neither necessary nor sufficient for proof status—but it is a desirable property, nonetheless.

An argument might convince a neutral audience, but fail to persuade a determined sceptic. Just this happened to Andrew Wiles's first attempt at a proof of the Fermat conjecture: the initial audience were convinced, but the argument ran into trouble when exposed to determined criticism from its referees. Such a case might be seen as success within an inquiry proof dialogue, followed by failure in a persuasion proof dialogue. A similar story could be told about Kempe's 'proof' of the four colour conjecture: a result which received far less scrutiny than Wiles's work, and was thereby widely accepted for eleven years. On the other hand, if even the sceptics are convinced, then an open-minded audience should follow suit. Thus, on the conventional understanding of mathematical rigour, success within both inquiry and persuasion proof dialogues is necessary for an argument to count as a proof.

We saw in the last section how a variety of differently motivated departures from the prevailing standards of mathematical rigour may be understood as shifts to different types of proof dialogue. Indeed, one of Walton's principal concerns in his analysis of natural argumentation is the identification of shifts from one type of dialogue to another. Such shifts can take a variety of forms: either gradual or abrupt, and either replacing the former type of dialogue or embedding the new type within the old. These processes are an essential and productive aspect of argumentation, but they are also open to abuse. Similar warnings apply to shifts towards less rigorous types of proof dialogue.

Many of the concerns which the critics of these forms of argumentation have advanced may be understood as an anxiety about illicit shifts of proof dialogue type. For example, the published discussion of mathematical conjecture is something which Jaffe and Quinn welcome: their concern is that such material not be mistaken for theorem-proving. Although some of their critics interpreted their advocacy of 'theoretical mathematics' as a

12 For example, von Staudt's proof of the equivalence of analytic and synthetic projective geometry has retained its difficulty for nearly two centuries. See [15, pp. 193 f.] for a discussion.

radical move, their primary goal was a conservative one: to maintain a sharp demarcation between rigorous and speculative work. Their 'measures to ensure "truth in advertising" ' [10, p. 10] are precisely calculated to prevent illicit shifts between inquiry and deliberation proof dialogues.[13] A similar story could be told about Tymoczko or Swart's discussion of methods they see as falling short of conventional rigour. Zeilberger is advocating the abandonment of rigour, but he recognizes at least a temporary imperative to separate rigorous from 'semi-rigorous' mathematics.

As Toulmin *& al.* remark 'it has never been customary for philosophers to pay much attention to the *rhetoric* of mathematical debate' [19, p. 89]. The goal of this article has been to exhibit some of the benefits that may accrue from a similarly uncustomary interest in the *dialectic* of mathematical debate—a dialectic which informal logic can do much to illuminate.

REFERENCES

[1] Jesús Alcolea Banegas, 1998, L'argumentació en matemàtiques, in *XIIè Congrés Valencià de Filosofia* (E. Casaban i Moya, editor), València, pp. 135–147.

[2] Kenneth Appel & Wolfgang Haken, 1978, The four colour problem, reprinted 2002 in *The philosophy of mathematics: An anthology* (D. Jacquette, editor), Blackwell, Oxford, pp. 193–208.

[3] J. Borwein, P. Borwein, R. Girgensohn & S. Parnes, 1995, *Experimental mathematics: A discussion*, CECM preprint no. 95:032, http://www.cecm.sfu.ca/preprints/1995pp.html.

[4] Wayne Brockriede & Douglas Ehninger, 1960, Toulmin on argument: An interpretation and application. *Quarterly journal of speech*, 46, pp. 44–53.

[5] Winston Churchill, 1937, *Great contemporaries*, Thornton Butterworth, London.

[6] Frans van Eemeren & Rob Grootendorst, 1992, *Argumentation, communication and fallacies*, Lawrence Erlbaum Associates, Hillsdale, N.J..

[7] Frans van Eemeren, Rob Grootendorst & Tjark Kruiger, 1987, *Handbook of argumentation theory: A critical survey of classical backgrounds and modern studies*, Foris, Dordrecht.

[8] Charles Hamblin, 1970, *Fallacies*, Methuen, London.

[9] Paul Hoffman, 1998, *The man who loved only numbers*, Fourth Estate, London.

[10] Arthur Jaffe & Frank Quinn, 1993, "Theoretical mathematics": Toward a cultural synthesis of mathematics and theoretical physics, *Bulletin of the American Mathematical Society*, 29, pp. 1–13.

[11] Alfred Kempe, 1879, On the geographical problem of the four colours, *American journal of mathematics*, 2, pp. 193–200.

[12] Donald MacKenzie, 1999, Slaying the Kraken: The sociohistory of a mathematical proof, *Social studies of science*, 29, pp. 7–60.

[13] Indeed, misleading advertisements are one of Walton's principal examples of an illicit dialogue shift. See, for example, [22, p. 206 ff.]

[13] Susan Newman & Catherine Marshall, 1992, *Pushing Toulmin too far: Learning from an argument representation scheme*, Xerox PARC technical report no. SSL-92-45, http://www.csdl.tamu.edu/~marshall/toulmin.pdf.

[14] Yehuda Rav, 1999, Why do we prove theorems?, *Philosophia Mathematica*, 7, pp. 5–41.

[15] Gian-Carlo Rota, 1997, The phenomenology of mathematical proof, *Synthese*, 111, pp. 183–196.

[16] Edward Swart, 1980, The philosophical implications of the four-color problem, *The American mathematical monthly*, 87, pp. 697–707.

[17] William Thurston, 1994, On proof and progress in mathematics, *Bulletin of the American Mathematical Society*, 30, pp. 161–171.

[18] Stephen Toulmin, 1958, *The uses of argument*, Cambridge University Press, Cambridge.

[19] Stephen Toulmin, Richard Rieke & Allan Janik, 1979, *An introduction to reasoning*, Macmillan, London.

[20] Thomas Tymoczko, 1979, The four-color problem and its philosophical significance, *Journal of philosophy*, 76, pp. 57–83.

[21] Douglas Walton, 1997, How can logic best be applied to arguments? *Logic journal of the IGPL*, 5, pp. 603–614.

[22] Douglas Walton, 1998, *The new dialectic: Conversational contexts of argument*, Toronto University Press, Toronto.

[23] Douglas Walton & Erik Krabbe, 1995, *Commitment in dialogue: Basic concepts of interpersonal reasoning*, SUNY Press, Albany, N.Y..

[24] Charles Willard, 1976, On the utility of descriptive diagrams for the analysis and criticism of arguments, *Communication monographs*, 43, pp. 308–319.

[25] Robin Wilson, 2002, *Four colours suffice: How the map problem was solved*, Allen Lane, London.

[26] Ludwig Wittgenstein, 1953, *Philosophical Investigations*, Blackwell, Oxford.

[27] Doron Zeilberger, 1993, Theorems for a price: Tomorrow's semi-rigorous mathematical culture, *Notices of the American Mathematical Society*, 40, pp. 978–981.

III

THE SPECIAL CASE OF MATHEMATICS EDUCATION

Chapter 9

MATHEMATICIANS' NARRATIVES ABOUT MATHEMATICS

and their relationship to its learning

Leone Burton
The University of Cambridge

Abstract: A study of the epistemologies of practising research mathematicians provides data with respect to the imaginative narratives (Bruner, 1986) used by these mathematicians when reflecting on their research practices. Unlike the paradigmatic narratives of formal mathematics, imaginative narratives involve the members of the mathematical community in active engagement, collaboratively, together with an acknowledgement of the holistic nature of knowing, thinking and feeling. The theoretical distinction is drawn between a contingent repertoire used within the imaginative mode when researching and the objectivist repertoire used to situate mathematics publicly as impersonal, separate and independent of the human or social. Focusing on the mathematicians' narratives, attention is drawn to how the mathematicians use transcendental or operational functions to support a simple objectivist stance and this is compared with the complexity of the contingent view couched in metaphoric and analogous reasoning. The chapter ends with a discussion of the implications of this analysis for mathematics education.

Key words: Narratives, objectivism in mathematics, contingent discourse, mathematics education.

1. INTRODUCTION

In this paper, I make the claim that, when learning mathematics, competency and comfort in the use of some particular formal mathematics is only acquired at the end of a complex chain of learning events. This chain commences with the telling by learners, and continues with the negotiation and re-negotiation between learners, of imaginative narratives about a mathematics problem. I described the imaginative mode as one "that inserts

B. Van Kerkhove and J.P. van Bendegem (eds.), Perspectives on Mathematical Practices, 155–173.

generality into the particularities of the narrative, attempting to tell engaging and believable stories which become exemplifications" (Burton, 1999d: 21). Only after experience of working at the different meanings ascribed to imaginative narratives by members of the learning community do learners subsequently arrive at a formal and recognisable statement of paradigmatic mathematics narrative, which seeks "to establish generalities out of particular examples, and then abandon[s] the particular in favour of the relentless drive of the logic of the general" (Ibid: 21). (The terms 'imaginative' and 'paradigmatic' derive from Bruner, 1986. See below).

My argument is built upon the content of the conversations that I held with practising research mathematicians in the course of a study of their epistemologies (see Burton, 2004). Why do I extrapolate from the learning experiences of mathematicians to learning experiences generally? First, researchers *are* learners and it seemed reasonable to assume that their learning processes might have something in common with less sophisticated learners of mathematics. Second, as mathematicians, they wield considerable power to influence future teachers through the learning practices that they make available in their undergraduate courses. Third, my observations of a gap between researching and teaching practices were confirmed in the study. This raised an important challenge to the continuation of teaching practices that are recorded as being ineffective. I am, therefore, making a case for providing learners with experiences, as described by mathematicians, that match their processes of researching mathematics (see Burton, 2004, 2001b, 1999a, 1999b, 1999c). The outcome of such a shift would be that learners, as much as researchers, would have access to the excitement and pleasure of engaging with mathematical enquiry. I would expect the result to be increasingly meaningful argument, and eventually the construction of paradigmatic narrative in mathematics.

This paper deals with the narratives told by mathematicians about the mathematics that they research. However, I am interested in narrative as the means by which we come to know. This is unlike Robert Thomas (2000, 2002, this volume) whose interests lies in comparing the narratives of mathematical products such as publicly produced mathematics texts with their similarities to and differences from fiction. Narratives reside within a culture and "a culture is as much a *forum* for negotiating and re-negotiating meaning and for explicating action as it is a set of rules of specifications for action" (Bruner, 1986: 123, original emphasis). Michael Toolan, writing of science that, in the perspective I am taking, can serve as a similar disciplinary area to mathematics, said:

> Science, too, may at first glance look very different from narrative. We often think of it as an expanding storehouse of incontestable facts, the

> hallowed repository of objective knowledge of how things in the world work: a rich but static description, quite remote from 'storytelling'. But that turns out to be mistaken in both theory and practice. (1988: xiv)

My focus, then, is on the stories, the imaginative narratives, that learners tell in their process of negotiation and re-negotiation of meaning, and in highlighting the ways in which these stories deviate from the 'paradigmatic' narratives (Bruner, 1986: 11-13) to be found in mathematics literature and, in particular, in texts. Jerome Bruner (1986) explained the meaning he gave to 'paradigmatic':

> One mode, the paradigmatic or logico-scientific one, attempts to fulfil the ideal of a formal, mathematical system of description and explanation. (Ibid: 12).

> [It] seeks to transcend the particular by higher and higher reaching for abstraction, and in the end disclaims in principle any explanatory value at all where the particular is concerned (Ibid: 13).

Paradigmatic mathematics is transcendent narrative in the sense that it is not realisable in experience, it is decontextualised, surpassing 'dirty' reality by employing the abstract. "The desire is for no less than that the grandeur and imprimatur of eternity be stamped on the objects of mathematics and the truths one discovers. In this way one can identify with a transcendent being" (Rotman, 1993: 156). It is unfortunately the case that the experience with mathematics of most learners is almost entirely at the level of the paradigmatic presented as received knowledge. This is consistent with Brian Rotman's observation that:

> the desire for the transcendental and its answering fantasy of mathematical absolutism have not been repudiated, resisted, or denied credence. It still holds mathematicians, and hence the rest of us, within its horizon, offering a paradisal theme-park of endless counting and an infinity of infinities without end – a play of pure unattached/disengaged signifiers within a discourse that achieves its sense, reference, and presence by severing any connection to any actual, potential, or imaginable corporeality (1993: 158).

All the more surprising, then, to discover, as I did, that the informal discourse employed by mathematicians, while not rejecting absolutism for the products of their research, nonetheless was one which *did* put human beings with all their fallibilities, and the community of which they are members, into the process of enquiry.

I am, therefore, concerned to explore the meanings available to learners, and their teachers within the "imaginative" narrative mode (Bruner, 1986: 13-14) as they refine and reform the shared meanings within their community of learners until, possibly only later, the narrative becomes a formal, paradigmatic text. Jerome Bruner explained the imaginative mode as dealing in:

> two landscapes simultaneously. One is the landscape of action, where the constituents are the arguments of action…The other landscape is the landscape of consciousness: what those involved in the action know, think, or feel, or do not know, think or feel (Ibid: 14).

So, building imaginative narratives involves the members of the community in active engagement, collaboratively, together with acknowledgement of the holistic nature of knowing, thinking and feeling. (For details on how these played out in the conversations with the mathematicians, see Burton, 2004.) This causes me to question the place in curricula given to, and the need for, universal published transcendent kinds of narratives written as paradigmatic mathematics, outcomes of which are their emphasis on the role of the individual and the product, mathematics, as the supreme triumph of cognition over emotion. The alternative, which attends to the demands of learning, as opposed to teaching, or at the least recognises a very necessary link in the chain of mathematical production, is provided by the imaginative narratives spoken by and with the human agents of a much more contextualised and holistic mathematics.

I begin by discussing the theoretical context within which the argument is placed. I then examine the mathematicians' statements about mathematics using the distinction between contingent and objective repertoires and between imaginative and paradigmatic narratives. (The meaning ascribed to these terms is explained above and below.) This discussion is divided between positions that are apparently straightforward and those that are complex. Finally I draw out the implications of this for mathematics education.

2. THEORETICAL CONTEXT

Mathematicians present their discipline as objective, certain, independent of the person, in every sense 'pure'. Unfortunately, or fortunately, depending upon your perspective, this picture is largely fiction. In this paper, drawing on data from a study completed in 1997, I demonstrate how mathematicians inhabit two conflicting worlds. In one, the world of their research, they are engaged in building narratives and, in discussing this world, they rely largely on what Gilbert and Mulkay (1984) have described as a "contingent repertoire" in which their accounts of mathematics are discursive and people-dependent, utilising the imaginative mode. When talking about mathematics the discipline, and especially when discussing teaching of mathematics which is almost universally perceived as requiring transmission of knowledge objects in a paradigmatic narrative mode, they usually resort to an objectivist repertoire in which they draw on beliefs about mathematics being either discovered or invented to convey a sense of its impersonality, separateness, independence from the human or social.

In Burton (1995), I described an epistemological model of knowing mathematics built from the philosophical and sociological literature (see, for example, Bloor, 1991; Davis & Hersh, 1983; Harding, 1986, 1991; Keller, 1985; Lakatos, 1983; Polanyi & Prosch, 1975; Restivo, 1992, 1985; Restivo et al., 1993; Rose, 1994; Rosser, 1990.) The model was not concerned with knowledge acquisition but with the process of coming to know, or knowing. It described knowing in mathematics in terms of five interacting categories:

- person- and cultural/social-relatedness,
- the aesthetics it invokes,
- its nurturing of intuition and insight,
- its recognition of different approaches particularly in thinking styles, and
- its connectivities.

In other words, I was conjecturing that mathematical knowing is a product of people and societies, that it is hetero- not homogeneous, that it is inter-dependent with feelings especially those attached to aesthetics, that it is intuitive and that it inter-connects in networks. This differentiates my focus from a knowledge-based enquiry.

The model proved to be remarkably robust when matched with how seventy research mathematicians, 35 women and 35 men, in universities in England, Scotland and Ireland, spoke, in an interview-based study, about how they come to know mathematics. They freely invoked the categories of the model. However, at the same time, they positioned themselves very differently along continua from positive to negative with respect to aesthetics and intuition and whether or not mathematics is a cultural artefact. I began the study with the expectation that the process of socialising a research mathematician as a member of that community of practice (Wenger, 1998) would be so robust that gender would not be influential as a distinguishing characteristic. That is, my expectation, which was fulfilled, was that the range of beliefs and attitudes about coming to know mathematics would be represented across women and men irrespective of gender.

Etienne Wenger (1998) explored the relationship between learning and negotiation of meaning, participation and reification, knowledge and knowing. The process he invoked was one in which participation in a community of practices involves the learner in negotiation of meaning and consequent knowing. This may lead to subject reifications, abstractions of the mathematical world such as symbols or tools that, themselves, become independently presented in the form of knowledge and are then utilised as objects of further meaning negotiation. He described living as "a constant process of *negotiation of meaning*" (Ibid: 53, original italics), the purpose of which is to learn. The negotiation of meaning involves "the interaction of two constituent processes … *participation* and *reification*" (Ibid: 52, original italics) and "their complementarity reflects the inherent duality of this process" (Ibid: 66). The relationship of participation and reification was:

> on the one hand, an intense involvement with the reificative formalisms of [the] discipline; and on the other, a deep participative intuition of what those formalisms are about (Ibid: 67).

He pointed out that:

> it is the interplay of participation and reification that makes people and things what they are. In this interplay, our experience and our world shape each other through a reciprocal relation that goes to the very essence of who we are…the river only carves and the mountain only guides, yet in their interaction, the carving becomes the guiding and the guiding becomes the carving (Ibid: 70/71).

Knowledge was "a matter of competence with respect to valued enterprises" and knowing, "participating in the pursuit of such enterprises" (Ibid: 4).

My concerns in this study were epistemological. I was focussing upon how mathematicians understood their own knowing and the process of constructing it. Hence, participation and reification were fundamental to my analysis of how they come to know and the ways in which they contribute to the negotiation of meaning. Equally important was the other interacting duality of the social and the individual represented in the practices of a community and the formation of and influence on its members' identities. (For a useful similar approach to scientists' discourse, see Gilbert & Mulkay, 1984.) So, far from being able to describe or predict coming to know objectively, I was exploring this process as one that is essentially human, and consequently socio-cultural.

In adopting this stance, I was surprised to find that just under 10% of the mathematicians I interviewed (6 out of 70) asserted that mathematics is an artefact of the socio-cultural. The number might be small, but it was far greater than I had reason to expect from speaking to mathematicians and reading the literature. Those who spoke in these terms, were articulate about why[1]:

> *I think mathematics is a product of people. It does turn out to be extremely useful. But even utility is a cultural product.*
>
> *The mathematics that we know, write down, is made up by men and women. Statements which have in today's society, been accepted as proofs, under the conditions of today's society, are what I am calling mathematical truths.*
>
> *Mathematics is a language, a way of communicating to the world just like hieroglyphics was to the Egyptians.*

The next mathematician raised a number of issues:

> *My view about mathematics is that it is a human invention and therefore is an intrinsic part of culture. I don't think that the mathematics classifies itself even though it looks as if it does. I don't believe that it does arrange itself into these kinds of hierarchies which appear to be intrinsic. People who look at mathematics sometimes reclassify it. It isn't the mathematics who does that, it is the people.*

His position was explained by Brian Rotman in the following way:

> mathematical signs do not fall from the sky, ready-made and pre-adapted to fit a world that knows nothing of them; they have themselves been extracted from whatever world mathematics has found and changed over the past several thousand years [...]Not only are its overall history, its range of particular concerns, and the direction and development of its research programs influenced by numerous social, material, and psychological factors external to itself [...] but the actual content and logical shape of mathematical signifieds themselves owe their origin to empirical, material features of the world (Rotman, 1993: 141).

The quotes from the above mathematicians make clear the potential for a clash such as I have proposed above, between objectivity and contingency. Their discourse was highly contingent and yet these same mathematicians, in their teaching, adopted an objectivist stance and, in their references to students, could be quite dismissive of the potential and motivation to engage with the shifting world that is described in these quotes. It would seem that, in order to live with the discrepancies between the two paradigms,

1 Direct quotations from the interviews are in italics.

mathematicians have to perceive research as a practice distinctive from, say, teaching. One consequence of this is that there is no motivation to reflect on their own learning in order to use the products of such reflection with their students. On the contrary, as Abbe Herzig showed, faculty members tend to position their Ph.D.students as needing to demonstrate "if they can cut it or not, and to force them to absorb and synthesize a large amount of information certifying that they have breadth of knowledge in mathematics" (2002: 189).

Discourse is central to issues of enculturation. Stephen Ball observed that discourse determines not only "what can be said, and thought, but also …who can speak, when, where and with what authority " (1994: 21). Although it is not the only aspect, because of its importance it is discourse on which I focus here. In this paper, we hear how mathematicians understand mathematics in quotes drawn from the database of the 70 interviews.

Bearing in mind my contention above that mathematicians are trapped between the objectivist and the contingent discourses, the distinction between participation and reification helps us to understand how this operates. Mathematics itself, as a body of knowledge and skills, is seen as objective and the reifications that are the product of the work of mathematicians are understood by many, especially teachers, as knowledge objects although Thomas (2002) points out the inappropriateness of understanding mathematics in this way. In their writing, mathematicians may allow such knowledge objects to be given 'life' in that they attribute human behaviours to the objects. Two examples of such writing are: "The prolonged good behaviour of the Hartree-Fock type wave function…" or "The functors H and K induce a duality…". Writing in this style is not particular to mathematics (see Gilbert and Mulkay, 1984). In discussing what they term a Truth Will Out Device (TWOD), Gilbert and Mulkay describe the objectivist use of the TWOD, part of presentation of reified mathematics, and compare this with the contingent discourse in a participatory environment:

> the fully-fledged TWOD, and the associated empiricist perspective, is likely to be used in those typical situations where the speaker is reconstructing events in a way which directly displays the correctness of his current scientific views. The TWOD is less likely to be appropriate in cases where the speaker is engaged in making his errors understandable and scientifically acceptable. In the latter situation, it seems that a more effective reconstruction of the speaker's actions can be achieved by setting them within a strongly contingent portrayal of scientific action and belief (Gilbert & Mulkay, 1984: 109).

An article written for presentation to the research community certainly constitutes such a 'reconstruction' situation and elsewhere (Burton & Morgan, 2000) the narrative devices that mathematicians use to ensure that they retain control over the objective status of their work have been explored. For example, papers are either written in the passive, third person or utilise the authorial "we", even where the paper is single-authored (see Burton & Morgan, 2000.) As Robert Thomas confirms, such writings are still narratives: "Between narrative and mathematics, the important similarity [...] is how they exemplify similar successful ways to give appropriate discussion to relations" (2000: 125). In my view, their stories are narratives because they are permeated by the perspectives of their authors; but in their published form, they are of a different quality and purpose (as dictated by the rules of the mathematical writing game) to the more informal and contingent discourse used about their research by the participants in the interviews. However, I am pointing out in this paper that the global term "mathematics" actually refers to the specifics of published mathematical writings and that there is considerable informal mathematical story-telling employed long before the paradigmatic writing is judged as ready to appear publicly. It is the process of proceeding through re-negotiations of the informal prior to shifting to the formal that, I believe, contributes to, and is constitutive of, learning.

Asking the participants in this study to explain to me their perspectives on what *is* mathematics, called out some objectivist statements. One way of understanding the drive towards objectivism is the degree to which it helps the mathematician to remain at a comparatively straightforward, uncomplicated level of discourse when talking about mathematics. (Although my concern here is not mathematical writing, as such, we must recognise the extreme, culturally dictated, constraints that also operate on writing style and formulation to ensure that they conform with objectivist requirements.) Once complexity is acknowledged, the messiness of human participation becomes inevitable and the discourse becomes contingent. This can be seen in the two sections below.

3. WHAT IS MATHEMATICS? - THE APPARENTLY STRAIGHTFORWARD VIEW

3.1 Pure and Applied Mathematics

The following statement was made by one of the participants in my study:

Mathematics is a very eloquent, beautiful subject in that it can make order out of chaos and has applications to virtually every field. It is a great simplifier. It is a very precise and eloquent way of stating things. The whole thing has a beauty of systems and symmetry, lack of symmetry even.

This is the kind of transcendent narrative of mathematical products that bears little correspondence with either researchers' or learners' experiences but does contribute to a social consensus on the nature of the discipline. Mathematics, or more precisely pure mathematics, is understood to simplify, to be widely applicable, precise and beautiful. In other words, it transcends the human and exists in spheres that are almost other-worldly. Such an argument helps to position the mathematician as special, indeed almost godlike. In this way, the elitism and hierarchy, implicitly and sometimes explicitly part of the discourse and behaviours within the mathematics community (see Burton, 2004, Herzig, 2002), are sustained; and they are incorporated into the participatory behaviours within the community. The straightforwardness of the initial transcendent story thereby contributes to the maintenance of community practices that are exclusive and about the exercise of power.

The consensus within the mathematical community about mathematics exists within a social context that recognises the difficulty experienced by most people in learning the subject; they fail to appreciate the very aspects that mathematicians celebrate, its power and its beauty; nor do most non-mathematicians comprehend whether, and if so how, power and beauty might contribute explicitly, as motivators, to mathematical work. So, by sustaining their setting apart from the rest of the community, mathematicians retain their own special-ness, and that of mathematics.

The gap between pure mathematicians such as the one quoted above, and applied mathematicians, is very wide. The applied mathematicians tended to dismiss what they described as the pure mathematician's engagement with form and beauty, while themselves being more concerned with problem solving in the 'real' world, making something happen, being useful.

Mathematics is, for me, understanding structures, how to solve problems which have been modelled mathematically.

Mathematics is useful, you can do things with it, you can model real things, you can make predictions, you can compare with experiments.

Mathematics itself I think of as being made up of concepts and ideas so I think of it as a very meaningful thing

The narratives of the applied mathematicians are operational – about what works. They are practical, of and in the 'real' world. Very different from transcendent stories, but equally straightforward, the applied mathematicians are engaged in stories that are about creating solutions.

3.2 Use of Metaphor

A focus on form, structure and relationships does not, of course, preclude metaphoric thinking or speaking. Metaphor was frequently invoked by the mathematicians, pure and applied, to describe mathematics and/or their process of coming to know. They seemed to rely on only two metaphors, the jigsaw puzzle and the journey. In narrative terms, the journey is a story of a quest whereas the jigsaw puzzle metaphor constitutes a completion story. (Newell and Swan, 1999 pointed out that the jigsaw metaphor is connected to a 'traditional', positivist, approach.)

How does a mathematician know when s/he knows?

> *It is like putting the right pieces into a jigsaw. You know you have the picture right. There is a right picture and quite a lot of what I have done over the years is essentially finding nice jigsaw puzzles for things that sometimes other people have done special cases of and not seen what the main jigsaw puzzle was.*

The objectivist repertoire is present here in the dominance of *right*: *the right pieces into a jigsaw, have the picture right, There is a right picture.* There is also just a flavour of self-congratulation in the dismissal of those who *have not seen what the main jigsaw puzzle was.*

While still presenting mathematics in a highly formal way, the next mathematician engages in lyrical description of her experience as a journey:

> *Maths is like mountains in more ways than one. If you are walking up a mountain in a fog and you keep on and it is jolly hard work and you struggle and you get to the top and suddenly the fog clears and you feel really good and you have this fantastic view. That's just like when you have suddenly discovered something. Perhaps you are on the top of the mountain and you can see very clearly the terrain and where you have to get to but you can't see all the little paths in the maze and you have to come down the mountain and you have to do work. And you get stuck in bogs and forests and all kinds of things but you know where you are going. I think that is what it is like.*

Again, the objectivist repertoire can be observed in *see very clearly the terrain and where you have to get to.* This mathematician invoked the public discourse of beauty in the reference to *this fantastic view.* At the same time, despite a brief reference to the involvement of the person in *hard work* and getting *stuck*, there is a sense of power in knowing *where you are going.*

I could imagine any one of the above mathematicians being content with a description of the discipline which emphasises its objectivity, independence from human activity, makes clear that it *is* its reificative formalisms. Such a story enables the mathematician to remain separate from the difficulties of living and to enjoy an intellectual life with their mathematics that is socio-culturally less demanding. It also encourages use of the very pleasing personal language of beauty and elegance that is recognisably part of public mathematics discourse.

The apparent simplicity of the above positions, therefore, does hide layers of more complicated relationships between the overtly simple, transcendent statements about mathematics, the solution-type statements of the applied mathematicians and the contingent repertoire they all used to explain their meanings. However, there were mathematicians who, from the outset, perceived complexity both in the mathematics and how it is derived and used and who struggled to place the discipline in a relationship with other socio-cultural inventions or artefacts and, consequently, with the participative practices exercised by the community.

4. WHAT IS MATHEMATICS? - THE COMPLEX VIEW

> *Mathematics is a discipline for everyone and everyone is a mathematician whether they realise it or not. There are difficult things in mathematics that, for a variety of reasons, we might never understand. It is a tool, it is a language, we use it for describing and talking about phenomena and it is an entity in itself, it has internal coherence, constituents. I think mathematics is abstract but it is so intimately related to nature that it is difficult to separate the two.*

This mathematician abandoned simple description rooted in formal devices and embarked upon a much more layered view of the discipline invoking the notion of mathematics not only as a *tool*, but also as a *language*. While there was an acknowledgement of the existence of reificative formalisms in the reference to *it is an entity in itself...* there was also a recognition of the availability, and potential power of mathematics to be used by people in their normal pursuits. This is particularly noticeable in the links drawn between *abstract* and *related to nature* so that *it is difficult to separate the two.* This is a contingent story that potentially can engage the mathematician in complications more related to socio-cultural issues than to the purity of the discipline.

Other mathematicians invoked the two discourse repertoires directly. For example, one registered a critique of the formalisms that make the discipline recognisable to those inside, and outside, of it, making use, again, of the journey metaphor. This time it was about a map, although a virtual not an actual map:

> *The definition, theorem, proof style is sometimes necessary to the health of mathematics - it can be over-prescriptive. People think that is what maths is, whereas I think it is about filling in gaps, making the map. Maths isn't what ends up on the page. Maths is what happens in your head.*

Another juxtaposed the simplicity of the model of certainty in mathematics with the challenge of the, not always provable, argument; to be convincing, however, the argument was still likely to be couched in objective terms:

> *I think it is very common for people not to question mathematics. What is there is true absolutely. When you get further in, we find that everybody is standing on the top of everybody else's shoulders, and there is no uncertainty. Then the bombshell falls. It doesn't now matter to me that it isn't certain. It is the challenge of building the tower not whether or not the tower will fall. You have theorems which you can neither prove nor disprove...it is very satisfying to know that I have something that I can tell you about and if you will listen to me I should be able to convince you. You should be able to understand, if you accept my foundation. Mathematics is like that. Life is not like that!*

Davis and Hersh (1983) discussed the difference between analytic (formal) solutions to a problem and those derived through the use of analogy. They said: "Analog mathematizing is sometimes easy, can be accomplished rapidly, and may make use of none, or very few, of the abstract symbolic structures of 'school' mathematics" (p.302); that is, the narrative is more likely to be of the imaginative form couched contingently. On the other hand, in analytic mathematics, "the symbolic material predominates" (p.303) as reifications in a paradigmatic mode. One of their examples was the "**Theorem**. *It is impossible to fill up a circle C with a finite number of nonoverlapping smaller circles contained in C"* (p.309, original emphasis). They pointed out that the analog solution was "visually obvious" (p.310) and for an analytical solution referred to Davis, 1965 with the comment: "The analog solution is so apparent that to insist on more is a piece of mathematical pedantry" (p.310). Nonetheless, we have all had the experience of being in a mathematical audience where a proof that would qualify as visually obvious is dismissed as "simply a demonstration". This is a further example of the domination of the public discourse by the paradigmatic, objectivist mode, "the desire to ground mathematics, once and for all, in something fixed, totally certain, timeless and prelinguistic" (Rotman, 2000: 58), a desire that fails to appreciate that "mathematics is not a building …but a process: an ongoing, open-ended, highly controlled, and specific form of written intersubjectivity" (Ibid: 58/59).

5. CONCLUSIONS - IMPLICATIONS FOR MATHEMATICS EDUCATION

The dependency model which, internationally, tends to control the ways in which mathematics is taught and learnt, is not one to which the following mathematician wanted to be subject when undertaking research:

For me, mathematics is very much a personal thing...It is about self growth. If you don't enjoy doing it and value it yourself, there isn't a lot of point in always looking for confirmation from somebody else.

However, given the comments about teaching made in the interviews, her teaching style was just as likely as that of her colleagues to be transmissive, operating on a body of knowledge principle in presenting paradigmatic narratives. The split between researching and teaching, or even more appropriately between researching and learning, was demonstrated in the following quote where the objectivist perspective dominated:

When you are a student in university you meet very beautiful theorems which are all the high points which are incredibly beautiful. And then you start doing research and you meet a lot of mathematics that is dirty. It is hard work and you have to see all the nasty bits and you have to do a lot of hard work before you get your elegant results.

Those discussions that I had with the mathematicians emphasised that they saw no dilemma about perceiving their experiences as learners and researchers as discrete and unconnected to their teaching. On the contrary, the beauty of the presented mathematics was assumed to be recognisable to the learner. The appreciation of that beauty was, apparently, expected without any guidance to learners about the cultural norms for recognising beauty, nor a discussion about the person-dependent nature of such a judgment.

> *Mathematics is like art: distinguishing between an elegant proof and a nice proof and saying that this is beautiful compared to the other not beautiful proof is aesthetics. What some people find beautiful is ugly for others.*

The lack of opportunity to pursue the derivative process, meet the dead-ends, tell the imaginative stories, deprives the learner of the potential for searching out and establishing meaning.

Nonetheless, the contingency of the process did not move the mathematicians too far away from the objective stance. Judgments are made, for example about beauty; and, despite much evidence to the contrary, jigsaw puzzles are discussed as if there is a 'last' piece: *The pleasure of doing mathematics is something like getting the last piece in a jigsaw puzzle. I do enjoy putting the pieces together.* However, the expression of feelings brings the human agency back into the mathematical field and clarifies the contradiction between learning paradigmatic mathematics, the derivation of which is dissociated from the learner, and being involved, with others, in an enquiry process which might be more complex, but is certainly more engaging.

I began this paper by asserting that mathematicians, like scientists, utilise two different discourse repertoires, the contingent and the objectivist and, of course, the stories that they tell are constrained by the norms of each discourse. Within the contingent, it is permissible to speak of individual reactions, of feelings, of error, of dead-ends and, indeed, of community practices which are dysfunctional:

> *Mathematics is full of hidden assumptions. You have to read a paper at least twice to get near the real agenda and then through a process of deduction to find out what on earth it is. Is it that the people who are doing it are blithely unconscious of the wider process or is it a deliberate process of mystification?*

But the objectivist discourse is not far away. Asked to describe what mathematics meant to him, one mathematician said:

> *Mathematics is that part of knowledge where structures can be abstracted and axiomatised in a completely systematic way without the recourse to examples.*

Why should this matter? It seems to me that the contingent discourse is a discourse of learning and the mathematicians, when researching, are behaving as learners. One of the strengths of the model put forward by Wenger (1998) is its emphasis on duality. While participation and reification are, together, the basis of generating knowing through the negotiation of meaning, they are inter-dependent. You cannot have participation, for example in dialogue, without the reified objects of words. Further, "participation is essential to repairing the potential misalignments inherent in reification" (Ibid. p.64) and vice versa. Under these circumstances, the derivation, existence and acquisition of such mathematical reifications as formal proofs, computer programs or theorems is only possible through participatory practices in the community.

The contingent repertoire is attached to a participatory process of enquiry through which imaginative narratives are constructed and de-constructed and meaning is re-negotiated. This constitutes the process of growth and discovery of which Davis and Hersh spoke:

> mathematics in process of growth and discovery, which is of course mathematics as it is known to mathematicians and students of mathematics. Formalized mathematics, to which most philosophizing has been devoted in recent years, is in fact hardly to be found anywhere on earth or in heaven outside the texts and journals of symbolic logic (1983: 347/8).

It is within this process that learning takes place. Associated with the process are not only collaborative community practices but also many of the behaviours for which learners call in mathematics classrooms and about the absence of which they despair (see Burton, 2001a, Boaler, 2002, Herzig, 2002). These are behaviours of inter-personal respect, opportunities for discussion and teamwork, responsibility for their learning and, above all, mathematical challenge. "It was the transmission of closed pieces of knowledge that formed the basis of much of the students' disaffection, misunderstandings, and underachievement" (Boaler, 2002: 183).

Within the mathematics community of practice, the politics around participation have not only led to practices such as, for example, third person writing being reified into game rules but, more seriously, to behaviours which are exclusive, hierarchical, competitive and dysfunctional (see Burton, 2004). It is my contention that such practices are closely tied to the objectivist, paradigmatic formulations that constitute the public voice of mathematics and that are then transferred into classrooms ensuring that mathematics is experienced negatively, and rejected by large numbers of students at every level across schools and universities, in undergraduate and postgraduate studies (see Burton, 2001a, Boaler, 2002, Herzig, 2002, Nardi & Steward, 2003). What many of the mathematicians, and students, ask for is participation built upon mutual engagement, joint enterprise and a shared repertoire which during learning is recognisably contingent. Using the contingent repertoire, the mathematicians reinforced the holistic nature of their enterprise. There is a consistency in the story here between contingency, imaginative narratives and a holistic approach as opposed to objectivism, paradigmatic narratives and the package of behaviours that maintain negative experiences of the discipline as described by so many.

However, recognising the relationships between knowing and knowledge as learning and as the outcome of research (see Brew, 1988), and understanding that this is independent of who is the learner, poses a serious challenge to those working in universities. It was apparent that many of the mathematicians with whom I spoke made no link between their learning from research, and their students' learning nor between their students' learning and their own practices as teachers of mathematics. Indeed, they did not make the links that I have been making in this paper between the practices of the community and their experiences of their discipline. And yet their epistemological processes as researchers, contradictory to the knowledge epistemology driving the discipline, were as appropriate to the learning of their own students and of less sophisticated and younger students as they were for their own purposes. I believe, therefore, that our failure to teach mathematics satisfactorily or successfully can be tied into the very issues that this paper has described. It is my belief that only when learning is understood as a result of research will it become possible to introduce and sustain effective practices amongst the community. And only at that time will the negative stories told by many in this study be recognised as dysfunctional, for mathematicians, students of mathematics *and* for the discipline.

By adopting a narrative perspective on mathematics, both on its participatory practices *and* on its reifications, I am challenging a long-held view on the subject. That this perspective exists in the contingent repertoire used by mathematicians to describe the processes of engaging with the discipline only confirms for me the importance of recognising it and raising it to equal status with the objectivist repertoire. That would then, it seems to me, enable a questioning of the efficacy of the objectivist repertoire together with recognition of when and where its importance for the discipline may lie.

ACKNOWLEDGEMENTS

I am very grateful to the following for their willing application of their critical reading skills and their helpful comments on this paper: Mary Barnes, Mary Coupland, Charlotte Franson and Christine Hockings.

REFERENCES

Ball, Stephen (1994) *Education Reform: A Critical and Post-structural Approach.* Buckingham: Open University Press.

Bloor, David (1991) *Knowledge and Social Imagery*, 2nd Ed., London: University of Chicago Press.

Boaler, Jo (2002) *Experiencing School Mathematics*, 2nd Edition, Mahwah NJ: Lawrence Erlbaum Associates.

Brew, Angela (1988) *Research as learning*, thesis submitted for the award of PhD by the University of Bath, UK.

Bruner, Jerome (1986) *Actual Minds, Possible Worlds*, London: Harvard University Press.

Burton, Leone (2004) *Mathematicians as Enquirers: Learning about learning mathematics.* Dordrecht: Kluwer/Springer.

Burton, Leone (2001a) Mathematics? No Thanks – Choosing and then Rejecting Mathematics. *Proceedings of a National Day Conference on Key Stage 3 mathematics teachers: the current situation, initiatives and visions*, pp. 58-71. The Open University, Milton Keynes UK.

Burton, Leone (2001b) Research Mathematicians as Learners – and what mathematics education can learn from them, *British Educational Research Journal*, **27**(5), 589-599.

Burton, Leone (1999a) Fables: The Tortoise? The Hare? The Mathematically Underachieving Male? *Gender and Education*, **11**(4), 413-426.

Burton, Leone (1999b) The Practices of Mathematicians: what do they tell us about Coming to Know Mathematics? *Educational Studies in Mathematics,* **37**, 121-143.

Burton, Leone (1999c) Why is Intuition so Important to Mathematicians but Missing from Mathematics Education? *For the Learning of Mathematics*, **19**(3), 27-32.

Burton, Leone (1999d) The Implications of a Narrative Approach to the Learning of Mathematics in L. Burton (Ed.) *Learning Mathematics: from Hierarchies to Networks*, London: Falmer Press.

Burton, Leone (1995) Moving Towards a Feminist Epistemology of Mathematics, *Educational Studies in Mathematics*, **28**(3), 275-291.

Burton, Leone & Morgan, Candia (2000) Mathematicians Writing. In *Journal for Research in Mathematics Education*, **31**(4), 429-453.

Davis, P.J. (1965) Simple Quadratures in the Complex Plane. *Pacific Journal of Mathematics*, **15**, 813-824.

Davis, P.J. & Hersh, R. (1983) *The Mathematical Experience*, Harmondsworth: Penguin.

Gilbert, G. Nigel & Mulkay, Michael (1984) *Opening Pandora's Box*, Cambridge: Cambridge University Press.

Harding, Sandra (1991) *Whose Science? Whose Knowledge?* Milton Keynes: Open University Press.

Harding, Sandra (1986) *The science question in feminism*, Milton Keynes: Open University Press.

Herzig, Abbe (2002) Where have all the students gone? Participation of doctoral students in authentic mathematical activity as a necessary condition for persistence toward the Ph.D. *Educational Studies in Mathematics*, **50**(2) 177-212.

Keller, Evelyn Fox (1985) *Reflections on gender and science*, New Haven CT: Yale University Press.

Lakatos, Imre (1983) *Mathematics, Science and Epistemology*, Cambridge: Cambridge University Press.

Newell, S. & Swan, J. (1999) Knowledge Articulation and Utilisation: Networks and the Creation of Expertise. Paper given at the User Workshop, *Knowledge Management and Innovation*, Royal Academy of Engineering, London, April 23.

Nardi, Elena & Steward, Susan (2003) Is Mathematics T.I.R.E.D.? A profile of quiet disaffection in the secondary mathematics classroom. *British Educational Research Journal*, **29**(3) 345-367.

Polanyi, M. & Prosch, H. (1975) *Meaning*, London: University of Chicago Press.

Restivo, Sal (1992) *Mathematics in Society and History*, Dordrecht: Kluwer.

Restivo, Sal (1985) *The Social Relations of Physics, Mysticism, and Mathematics*, Dordrecht: D. Reidel.

Restivo, S., van Bendegem, J.P.& Fischer, R. (Eds.) (1993) *Math Worlds: Philosophical and Social Studies of Mathematics and Mathematics Education*, Albany: State University of New York Press.

Rose, Hilary (1994*) Love, power and knowledge: towards a feminist transformation of the sciences*, Cambridge: Polity Press.

Rosser, S.V. (1990) *Female-Friendly Science*, New York: Pergamon.

Rotman, Brian (2000) *Mathematics as Sign: Writing, imagining, counting*, Stanford CA: Stanford University Press.

Rotman, Brian (1993) *Ad Infinitum: The Ghost in Turing's Machine – Taking God out of Mathematics and Putting the Body back in*, Stanford CA: Stanford University Press.

Thomas, R.S.D. (2002) The comparison of mathematics to narrative. Paper delivered to the Conference, *Perspectives on Mathematical Practices*, 24-26 October 2002, Brussels, Belgium.

Thomas, R.S.D. (2000) Mathematics and Fiction I: Identification. *Logique & Analyse*, **43**(171-172), 301-340.

Toolan, Michael J. (1988) *Narrative: A Critical Linguistic Introduction*, London: Routledge.

Wenger, Etienne (1998) *Communities of Practice*, Cambridge: Cambridge University Press.

Chapter 10

PHILOSOPHY OF MATHEMATICS AND MATHEMATICS EDUCATION

The Confluence of Mathematics and Mathematical Activity

Anthony Peressini[1] and Dominic Peressini[2]
[1] Marquette University; [2] University of Colorado at Boulder

Abstract: In this paper we explore how the naturalistic perspective in *philosophy of mathematics* and the situative perspective in *mathematics education,* while on one level are at odds, might be reconciled by paying attention to actual mathematical practice and activity. We begin by examining how each approaches mathematical knowledge, and then how mathematical practice manifest itself in these distinct research areas and gives rise to apparently contrary perspectives. Finally we argue for a deeper agreement and a reconciliation in the perspectives based on the different projects of justification and explanation in mathematics.

Key words: Mathematical practice, mathematics education, proof, mathematical ontology, mathematical epistemology, naturalism, situative learning theory

Researchers in the philosophy of mathematics and mathematics education both have a deep and natural interest in the actual practice of mathematics; despite this, the mainstream in each of these distinct disciplines seem at best to have little interchange and at worst, deep disagreement.

This paper is an exploration of ways in which two "mainstream" researchers in philosophy of mathematics and mathematics education, and their research speak to one another. In particular, it explores the different ways each theorize and utilize mathematical practice with respect to an issue common to both disciplines, namely, the nature of mathematical knowledge and its use of proof.

We begin the next two sections by examing how mathematical practice manifests itself in our distinct research on mathematical knowledge and

B. Van Kerkhove and J.P. van Bendegem (eds.), Perspectives on Mathematical Practices, 175–189.

proof (Peressini, A., 1999, 2003; Borko, H., Peressini, D., Romagnano, L., Knuth, E., Yorker, C., Wooley, C., Hovermill, J., & Masarik, K., 2000; Peressini, D., Borko, H., Romagnano, L., Yorker, C., Wooley, C., Hovermill, J., & Masarik, K, 2001; Peressini, D. & Knuth, E., 1998). This is followed by a presentation from an ongoing research project in mathematics education[1] that highlights the nature of proof in a variety of educational settings. Finally we present *prima facie* contrary conclusions from a mainstream philosophy of mathematics perspective followed by a brief discussion that attempts to reconcile these distinct perspectives.

1. PHILOSOPHICAL NATURALISM

Within the philosophy of mathematics (and philosophy in general), it is Williard Quine's (1969a, 1981a) idea of "epistemology naturalized" that most prominently urges the "looking to practice" to determine the answer to important philosophical questions such as what kind of things exist (ontological questions) and what and how we know (epistemological questions). This naturalistic approach eschews the excesses of the philosophical traditions that sought to answer such questions by *a priori* ruminations or "arm chair" philosophy; it refuses to fall into the arrogance of foundational approaches (like Descartes or Kant) that sought to "reform" the science of their time from a framework of "prior" philosophical assumptions.

Naturalism, most generally, asserts that in philosophy, we cannot do better than our best scientific account of the world for a *starting point*, and that philosophical accounts ought to *remain* "continuous" with scientific theory and practice. Naturally, this counsels us to look to science to understand the world — including the mathematical world.

Naturalism is founded on the belief that the role of philosophy is to help clarify and better understand our "ordinary talk of physical things." (Quine, 1960, p. 3) The natural starting place for such a project is with our best account of the world and our place in it, and this is to be found in our current scientific theories. A naturalized epistemology has no place for pre-scientific principles of knowledge; it recognizes no ground outside of

[1] The Learning to Teach Secondary Mathematics project (LTSM) is a multi-year research project funded in part by the National Science Foundation (NSF Grant #REC-9605030 and REC-0087653). We would like to acknowledge Hilda Borko and Lew Romagnano for their work on this aspect of the paper as they are Co-PIs with Dominic Peressini on the LTSM project.

science from which we can scrutinize, revise, and reconstruct knowledge of the world.

It is part of the job of the philosopher to criticize and help clarify the work of the scientist, but this must be done on *scientific grounds.*

As applied to the question mathematical ontology, Quine's naturalism gives us the Quine/Putnam Indispensability Argument (Quine 1969b, 1981b; Putnam 1971, 1979a) for a realistic stance with respect to mathematical objects. The idea is that we ought to believe in precisely those objects that our best scientific account of the world finds indispensable to its work, i.e., things like electrons, forces, genes, and *also* things like numbers, functions, sets, and the like. This naturalized approach has taken quite a hold with respect to ontological issues.

With respect to traditional epistemology, naturalism hasn't had as obvious a presence. The Quinean idea is that traditional epistemology ought to be grounded in cognitive psychology (and related scientific areas of inquiry) — so again we look to our best (scientific) account of how we know to understand what knowledge is.

The lack of influence of this aspect of Quine's naturalism is quite likely because of the "normative" component of knowledge. The normative component, on the traditional account is justification, and justification, quite naturally, (and rather famously) resists characterization in descriptive terms such as those of cognitive psychology. We'll speculate also (and elaborate below) that the narrow traditional focus on propositional knowledge (i.e., the knowledge *that* a proposition is true) that has dominated the discussion in the philosophy of mathematics has also played a role.

In an attempt to pursue the naturalized ideal of looking to our best account of a phenomenon as a way of (at least) informing our philosophical account it, why stop at cognitive psychology? In particular, when considering the nature of mathematical knowledge and how we know mathematically, shouldn't we look more broadly to the learning and educational theory surrounding mathematics? We think this approach does have, at least, initial merit, and our work together is really the project of investigating whether there is in fact deeper potential.

2. EDUCATION THEORY: A SITUATIVE PERSPECTIVE

For teachers, learning occurs in many situations of practice. These include university mathematics and teacher-preparation courses, preparatory field experiences, and schools of employment. A situative perspective argues that, to understand teacher learning, we must study it within these multiple contexts, taking into account both the individual teacher-learners and the physical and social systems in which they are participants.

Traditional cognitive perspectives typically treat knowing as the manipulation of symbols inside the mind of the individual. Learning is typically described as an individual's acquisition of knowledge, change in knowledge structures, or growth in conceptual understanding. Cognitive theorists argue that, while some learning takes place in a social context (e.g., on-the-job training), what is learned can also be independent of the context in which it is learned (Anderson et al., 1997).

Perhaps the most impressive consideration of teaching from a cognitive perspective is the work of Alan Schoenfeld and his Teacher Model Group to develop a theory of teaching-in-context (Schoenfeld, 1998). Schoenfeld and colleagues are attempting, through the use of cognitive modeling strategies, to explain teachers' decisions at each point of instruction by identifying the goals, beliefs, knowledge, and action plans on which those decisions are based.

In an invited critique of this work, Greeno (1998) suggested that a situative perspective would add significantly to this understanding of teacher decisions by focusing on classroom social practices and examining such features as the patterns of discourse, the kinds of participation that are afforded to the teacher and students by the classroom practices that are in place, and the personal identities developed by the teacher and students through participation in these practices.

We find a situative perspective to be compelling as a framework for the study of teacher learning across time and across the multiple contexts of early-career teaching. We conceptualize the novice teacher's learning-to-teach experiences as a single learning trajectory through the multiple contexts of teacher education. A situative perspective offers us a way of disentangling—without isolating—the complex contributions of these various contexts to novice teachers' development.

2.1 Proof and justification

Proof is central to the discipline of mathematics and the practice of mathematicians. Yet, its role in secondary school mathematics has traditionally been peripheral, usually limited to the domain of Euclidean geometry. Current reform efforts, however, call upon secondary mathematics teachers to provide all students with rich opportunities and experiences with proof throughout the secondary mathematics curricula. For example, in *Principles and Standards for School Mathematics* (NCTM, 2000) a separate standard on proof recommends that:

> Instructional programs from pre-kindergarten through grade 12 should enable all students to: recognize reasoning and proof as fundamental aspects of mathematics; make and investigate mathematical conjectures; develop and evaluate mathematical arguments and proofs; [and] select and use various types of reasoning and methods of proof (p. 56).

Are secondary school mathematics teachers prepared to enact these recommendations in their instructional practices? To date, little research has focused on secondary mathematics teachers' conceptions of proof. Researchers have focused primarily on students' conceptions of proof (Balacheff, 1991; Bell, 1976; Maher & Martino, 1996), prospective elementary school teachers' conceptions of proof (Martin & Harel, 1989; Simon & Blume, 1996), and undergraduate mathematics majors' conceptions of proof (Harel & Sowder, 1998). Further, much of the previous research has investigated individuals' understandings of proof and methods of proof. Our framework broadens this focus to include individuals' situated knowledge of the *social nature* of proof, and of the *role* of proof in establishing and explaining mathematical truths across a variety of contexts.

Many mathematicians and mathematics educators view proof as a *social process* engaged in by members of a community of mathematical practice. Proponents of this view describe proof as "a debating forum" (Davis, 1986, p. 352), "a form of discourse" (Wheeler, 1990, p. 3), "a social construct and a product of mathematical discourse" (Richards, 1991, p. 23), "a justification arising from social interactions" (Balacheff, 1991, p. 93), and "an essentially public activity" (Bell, 1976, p. 24). Consonant with this view of proof is the approach to mathematical growth and discovery outlined by Lakatos (1976):

> [M]athematics does not grow through a monotonous increase in the number of indubitably established theorems but through the incessant improvement of guesses by speculation and criticism, by the logic of proofs and refutations (p. 5).

By "proof," Lakatos is referring to the explanations, justifications, and elaborations that serve to make a conjecture more convincing and accurate in the face of refutations (i.e., counterexamples) posed by potential doubters.

Few would question that the main *role* of proof in mathematics is to demonstrate the correctness of a result or truth of a statement (Hanna, 1991). Yet, mathematicians expect the role of proof to include more than justification and verification of results: "mathematicians routinely distinguish proofs that merely demonstrate from proofs which explain" (Steiner, 1978, p. 135). Hanna (1990) elaborated on the distinction between these two types of proofs:

> A proof that proves [i.e., a proof that demonstrates] shows only *that* a theorem is true; it provides evidential reasons alone....A proof that explains, on the other hand, also shows *why* a theorem is true; it provides a set of reasons that derive from the phenomenon itself....[It] must provide a rationale based upon the mathematical ideas involved (p. 9).

Secondary mathematics students have traditionally conceived of proof as a formal, and often meaningless, exercise to be done for the teacher (Alibert, 1988). "In most instructional contexts proof has no personal meaning or explanatory power for students" (Schoenfeld, 1994, p. 75). However, mathematicians value a proof as much for its explanatory power as for its deductive mechanism. These two features of proof—explanatory power and deductive mechanism—address aspects of teachers' conceptions of proof that are essential for understanding their implementation (or lack thereof) of reform recommendations concerning proof in school mathematics.

3. EDUCATIONAL PRACTICE: WHAT IS A PROOF?

In fall 1995, Audrey Savant[2] put on hold her professional life as a musician and teacher of private flute lessons to obtain a secondary mathematics teaching license. Armed with two degrees in music, she attended the undergraduate-only Metropolitan College, where she completed a mathematics major (including reform-based courses in the foundations of geometry and mathematics teaching methods) and the teacher preparation program. In spring 1999 she student taught at Cumulus High School, a three-year-old school of 1,600 students in a rapidly-growing upper-middle class and predominately white suburb. Ms. Rockford, her cooperating teacher (an amateur musician herself), is a National Board for Professional Teaching Standards certified teacher and an instructional leader in the school district.

Our observations of Ms. Savant in several research contexts prior to student teaching revealed that she was quite adept at one of our mathematics domain slices: proof. We present three brief examples here. In one research task, part of an extended interview designed to elicit mathematics content knowledge, Ms. Savant was asked to prove the following:

Theorem: $1 + 2 + 3 + \ldots + n = n(n + 1)/2$.

She used the method of Mathematical Induction correctly to produce a proof. She was then asked to assess the validity of several different purported "proofs" of this theorem, including one that was almost identical

[2] The names of participants and their schools are pseudonyms.

to hers. She commented that the method of Mathematical Induction produces a valid proof "if you follow all of the steps correctly."

Our second example comes from Ms. Savant's Foundations of Geometry class, an upper-division requirement for her mathematics major. She and her classmates were continuing their semester-long project of building a set of axioms for Euclidean Geometry, what the professor described as "a minimal set of descriptors of what we think is good geometry." On this evening, the following question was on the table: As conditions for establishing triangle congruence, are both SAS and ASA correspondences[3] needed as axioms, or does the second follow as a consequence of the first? The professor stated the theorem: SAS implies ASA. He then said to the class that he would "get them started" on the proof. He wrote on the board, in very careful language and using a clearly labeled diagram, the first few steps of a proof by contradiction, in which two triangles were in ASA correspondence but were not congruent. Ms. Savant completed the proof at the board, carefully writing the steps that led to a contradiction. With the prodding of the professor, she completed her work by writing "Thus, we have a contradiction."

Our final example occurred in Ms. Savant's mathematics methods class, which she completed in the semester just before student-teaching. One evening, this class began with the following task on the overhead projector:

> Place a point A on a sheet of paper. Draw a line through A, and draw a point B about an inch from the line. Use the line you drew as a line of reflection, and use the MIRA to find the image of the point B. Use at least five other lines of reflection through A to find image points for B. If you were to find the set of images of B across all possible lines of reflection through A, what figure would this set of points form? Why?

After about 8 minutes of work in small groups, the instructor asked the class, "All right, so what kind of figure do you get?" Students quickly concluded that the figure is a circle centered at A with radius AB. The instructor asked, "How do you know?" Ms. Savant stepped to the board and offered a proof, in which she showed that all of the image points of B are the same distance from A, to justify the class' conclusion.

Given this emerging picture of Ms. Savant's knowledge of proof, we were drawn to a particular incident observed during her student teaching. In this class, Ms. Savant conducted an activity that she hoped would make an important connection for her students. In this activity, students cut paper circles into sectors and rearranged the sectors to form a rectangle-like figure. Ms. Savant then posed a series of questions about the figure, to help students connect the formula for the area of a rectangle to the formula for the area of a circle. She concluded by telling her students that they had just proved the formula for the area of a circle: "You cut up the area, you rearranged it, and you proved that the area [of a circle] is 'pi-r-squared.' That's great!"

We were interested in Ms. Savant's use of the word "proved" in this context. In the interview conducted later that day, the we asked Ms. Savant whether she thought the sectors-of-a-circle activity was a proof. In her response, Ms. Savant described proof as an informal sense-making process that shows why something is true.

> Yes I do [think this is a proof]. It's not a formal proof in that we sit down and we write it all out. But certainly it's a way to prove to yourself, "Hey look, this actually really does work. I can use things that I know and this really does work."

She commented further,

> "My idea of a formal proof is when you actually sit down, and you use symbolic language with English language, and you give either a paragraph or a step-by-step proof that's very logical."

Ms. Savant contrasted this notion of a formal proof with that of an informal proof, or convincing argument, which is acceptable and appropriate in a high school classroom.

Ms. Savant had made this distinction between formal and informal proofs often in interviews prior to this incident. She had repeatedly described formal proof as the structured symbolic presentations acceptable to mathematicians: "A proof is a logical and clear series of steps that take you from point A to point B." In contrast, for Ms. Savant informal proofs are ways to explain why something is true, particularly to students: "I think convincing [informal proof] is different from formal proof. But, in the vernacular English, it would prove to the kids that it works."

Through our situative lens, we see that Audrey Savant developed two somewhat different conceptions of proof as she progressed through her teacher education program—formal proofs that prove, and informal proofs that explain. She learned how to draw upon these conceptions in order to participate successfully in two different types of mathematical communities. In her content courses and on our research tasks, where the standards of proof were seen to be more related to structure and form than to explanatory power—proofs that prove, not necessarily explain—Ms. Savant drew upon a conception of proof that supported her successful participation as a student.

In her mathematics methods course and field placements—where the emphasis for her was on fostering student learning—she drew upon a very different conception of proof that contributed to her ways of participating as a teacher.

Ms. Savant's reliance on different conceptions of proof in different situations, viewed through a situative lens, has also helped us to understand why researchers have found little relationship between subject matter knowledge, as measured by the number of courses completed successfully, and teaching competence (Grossman, Wilson, & Shulman, 1989). The situations of content preparation—their norms and expectations and the roles played by individuals—are fundamentally different from the situations of teaching practice, and the patterns of participation learned in these two types of situations differ fundamentally as well. Successful participation as a student in undergraduate mathematics courses often requires use of several familiar proof *structures*, such as induction and proof by contradiction. On the other hand, participation as a teacher in a school mathematics classroom often consists of *explaining* and *justifying*, and providing students with opportunities to do the same. Successful participation in one of these situations does not necessarily support successful participation in the other.

4. PHILOSOPHICAL DIFFERENCES AND RECONCILIATION

What sort of lessons does the situative perspective on a case like this hold for a philosopher of mathematics? We'll focus on the one mentioned briefly above, namely, that perhaps philosophers ought to complicate a bit the conception of knowledge that we bring to our consideration of *mathematical* knowledge.

Many philosophers who work in the philosophy of mathematics are still in the grip of the old JTB theory of knowledge at some not insignificant level. It seems fair to say that the idea that *S* knows *p* iff *S* has a justified true belief that *p* still dominates epistemological thinking and for sure, teaching. Even clearer is that most philosophical discussions of knowledge are limited to *propositional* knowledge. I know I am not the only philosopher who, on the first day of an undergraduate epistemology class, distinguishes between real knowledge (propositional knowledge) and other non-epistemic senses of the work like acquaintance/object knowledge (I know Chicago or I know George Bush) and procedural knowledge (I know how to ride a bike or I know how to square a circle). And this focus is not limited to teaching: research in philosophy of mathematics has also focused rather exclusively on propositional conceptions of knowledge. Examples include work on the issues of the confirmation of a body of mathematical theory (Peressini, A. 1999) or work on the confirmation of a mathematical proposition relative to a body of mathematical theory (Peressini, A. 2003).

Our recent attention to mathematics education and education theory and its results like the one described above lead us to wonder if philosophers haven't been missing some things with this focus.

The case of Ms. Savant draws our attention to the dramatically different types of learning activities (and presumably knowledge) that are emphasized between elementary/high school and undergraduate/graduate mathematics education. The different senses of "proof" employed by Ms. Savant are, we believe, indicative of a different sense of mathematical knowledge at work in each context. In one context, the knowledge consists in a rather holistic collection of components — indicated by terms like skills, abilities, insights, conceptualizations, and recognitions. But the focus (at least as far as instruction goes) in the other, upper-level college context is on mathematical propositions as justified (by proof). A shift seems to occur; the educational goal seems no longer to be on developing the same mathematical ability, insight, and skills, but rather on exploring a body of theory unified by its justifying proof in a way chosen for its historical or genetic or disciplinary organizational structure — not in any obvious way primarily insight/learning goal driven.

Of course these contexts are not found in the exaggerated extremes just described. For example, an advanced (third year) calculus course taken in college is the first course emphasizing proof not problems, but it does typically work in a limited way for knowing in the more inclusive sense – at least one needs something like that kind of knowledge in order to do well on the exams. Note, however, that such instruction most often does not include anything obviously designed to *facilitate* this type of knowledge. The emphasis in the classroom is on justifying the true propositions of the mathematics under study and no doubt a desired outcome is the development of the limited proof skills necessary to engage in this sort of justifying activity. The question is why? It certainly is implausible to think of such courses as trying to prepare students for work as research mathematicians. More importantly for us here, how was it expected to come about? It seems clear that this only becomes more pronounced in graduate education, but it makes sense at this level, since a goal of a mathematics graduate program is to train research mathematicians. For the sake of time, we'll pass over problem-solving focused college courses such as business calculus and differential equations for engineers, though these pose interesting questions too.

Now an important question is whether this clear difference in educational methods/goals emphasized in these two contexts gives rise to or indicates importantly different kinds of mathematical knowledge. If it does, then this might make sense of the emphasis in philosophy of mathematics on propositional knowledge. Could it be that in fact mathematical research as done by professional mathematicians is largely a matter of justifying mathematical propositions? Certainly, an artful and complicated endeavor, and also one that tends to reward with successful individuals with the deep and holistic understanding that (good) mathematics educators try to instill in us, but couldn't it still, at bottom, have a primarily justificatory goal?

Notice that a justificatory account of the goal of mathematics is not necessarily incompatible with the (undeniable) recognition that in mathematics, some proofs are valued for their *explanatory* power. For it could just be that explanatory value is reducible to justificatory value; that is, explanatory proofs may be valued because of their utility in developing and justifying further theory. After all, isn't this the basis for explanatory value in science? On this account of mathematical practice, the work of mathematics (by mathematicians) is seen as developing our mathematical propositional knowledge, while the work of mathematics educators is seen as facilitating the incorporation of this propositional knowledge into our cognitive/psychological structures in rich ways that foster the use of the mathematical knowledge — both directly in application to the world and indirectly in developing further mathematical theory.

We are not sure, however, that things divide up so neatly. In order to sustain this defense of mathematical epistemology's rather exclusive focus on propositional knowledge, one has to buy into a neat division between the psychological/cognitive/subjective component, which is taken to be the belief that the proposition is true (i.e., a propositional attitude) and the objective normative component, namely the property the belief has of being justified or rationally or evidentially supported. But doesn't learning theory, which emphasizes the individual subjective nature of justification, militate against just this?

Maybe not. Here is where research in education and philosophy may butt heads. It certainly doesn't follow from the fact that the sense of justification relevant in the context of an individual's understanding is subjective that there is no epistemically normative sense of justification that is not essentially subjective. Put another way, just because our individual sense of justification is subjective *does not* entail that there isn't an objective sense of justification that is *normative* for our subjective sense, normative in that our subjective sense *ought* (rationally) to track it. This point is analogous to the following ethical version: that our individual sense of right/wrong is subjective *does not* entail that there isn't an objective sense of right/wrong that is *normative* for our subjective sense, normative in that our subjective sense *ought* (morally) to track it.

Intuitions may be screaming here that nonetheless there does seem to be a subjective aspect to justification, and even to the justificatory instrument *par excellence*, proof itself. Let us point to one further observation raised by the situative perspective in education theory that may assuage this concern. Notice specifically what it is that is objective in the account just offered — justification. We construe justification as a relationship, obtaining or not, between a proposition and a body of propositions. However, with respect to another important relationship, explanation, we follow von Fraassen (1980) and many others in taking it to be essentially pragmatic. The rough idea is that an explanation E explains a phenomenon P relative to a background B, where B is necessarily composed in part by pragmatic interests. When faced with a proof that we are inclined to label as "merely justificatory," a consequence of the pragmatic account of explanation is that such explanatory value (or lack) is not absolute in the way justification is; it is dependent on the relevant pragmatically determined background conditions. One then naturally wonders whether and what sort of background conditions would render such a proof explanatory. We offer a simple example to illustrate this point.

As is well known, mathematicians themselves distinguish between proofs that are explanatory and proofs that merely justify. (See for example Steiner 1978 or Rav 1999 for discussion and references.) As an example, consider the following elementary proof that 0.999... = 1.0.

(1)	$s = 0.999\ldots$	definition
(2)	$10s = 9.999\ldots$	multiplying both sides by 10
(3)	$9s = 9.0$	subtracting s from both sides and (1)
(4)	$s = 1.0$	dividing both sides by 9
(5)	$0.999\ldots = 1.0$	substitution and (1)

While this proof does establish that 0.999... = 1.0, it seems to offer nothing to explain why this equality holds. Such an explanation would probably involve considerably more, e.g., explaining the distinction between the rational numbers themselves and a decimal representation of them, how the decimal representation is related to a (potentially) infinite series, and also the Cauchy-Weierstrauss property (or an equivalent one). But this is not the end of the story. This simple proof may actually, in certain less obvious contexts, have explanatory value. For example, in an early arithmetic setting the proof might help explain why 0.33... <> 0.4, while 0.99... is equal to 1.0. Or alternatively, in the introductory algebra context of trying to develop a general method for finding the rational fraction associated with an infinitely repeating decimal, this proof may indeed be of explanatory value.

The interest-relativity of proof is present at the highest levels of mathematics as well. In preparing for research level work in analysis in top-tier graduate programs, students know that one must have both Walter Rudin's (1986) *Real and Complex Analysis* and H. L. Royden's (1988) *Real Analysis*. Rudin's presentation of real and complex analysis is so wonderfully elegant, concise, complete, and theoretically rich that it is basically useless to learn from for an inexperienced student; instead, students begin with Royden's presentation, which is much more plodding and direct. But as one moves along in a PhD program, past qualifying exams and into research, Rudin's proofs, which develop and made use of more generalized topological and algebraic proof machinery, become more explanatory than Royden's more plodding and direct versions.

5. CONCLUDING REMARKS

So it does seem that a reconciliation of epistemology's focus on propositional knowledge with learning theory's focus on holistic non-propositional knowledge can be started by carefully distinguishing between the distinct though related projects of justification and explanation. Indeed, such recognition makes clear why philosophers of mathematics stress the objectivity of mathematical knowledge, while learning theorists stress the subjectivity.

Philosophy's focus on propositional knowledge, if seen as addressing *exclusively* the justicatory problem can quite reasonably give rise to a deeply objective sense of knowledge. And this objectivity is quite compatible with learning theory's situative perspective, which at first glance to many philosophers of mathematics appears to entail the subjectivity of mathematical knowledge, if the situative perspective is understood as focusing not on propositional knowledge, but rather on knowledge of mathematical explanation.

This does then reopen questions pertaining to the relatively neglected areas of non-propositional knowledge. It would seem that the explanation knowledge of mathematics learning theory is something like the dismissed procedural knowledge of traditional epistemology. But how are they related? Is explanation knowledge a subset of procedural knowledge? Is it essentially performative or active? How does such a conception bolster or undercut the standard accounts of the social dimension of mathematical knowledge? Can attention to explanation knowledge reinvigorate Quine's naturalism in epistemology with its focus on cognitive psychology?

As we hope to have given you a taste of, we do see value in attempting to bring together mathematical practice as understood by naturalistic philosophers *and* mathematical practice as understood by educators from the perspective of a situative framework. We believe that the clearest indication of its value lies in its ability to reframe and contribute in an novel way to important neglected questions in both disciplines.

REFERENCES

Alibert, D. (1988). Towards new customs in the classroom. *For the Learning of mathematics, 8* (2), 31-35.

Anderson, J. R., Reder, L. M., & Simon, H. A. (1997). Situative versus cognitive perspectives: Form versus substance. *Educational Researcher, 26*(1), 18-21.

Balacheff, N. (1991). Treatment of refutations: Aspects of the complexity of a constructivist approach to mathematics learning. In E. von Glasersfeld (Ed.), *Radical constructivism in mathematics education* (pp. 89-110). Netherlands: Kluwer Academic Publishers.

Bell, A. W. (1976). A study of pupils' proof-explanations in mathematical situations. *Educational Studies in Mathematics, 7*, 23-40.

Borko, H., Peressini, D., Romagnano, L., Knuth, E., Yorker, C., Wooley, C., Hovermill, J., & Masarik, K. (2000). Teacher education does matter: A situative view of learning to teach secondary mathematics. *Educational Psychologist, 35*, 193-206.

Davis, P. (1986). The nature of proof. In M. Carss (Ed.), *Proceedings of the Fifth International Congress on Mathematical Education,* (pp. 352-358). Adelaide, South Australia: Unesco.

Greeno, J. G. (1998). Where is teaching? *Issues in Education: Contributions from Cognitive Psychology*, *4*(1), 111-119.

Grossman, P.L., Wilson, S.M., & Shulman, L.S. (1989). Teachers of substance: subject matter knowledge for teaching. In M. Reynolds (Ed.), *Knowledge base for the beginning teacher* (pp. 23-36). New York: Pergamon.

Hanna, G. (1990). Some pedagogical aspects of proof. *Interchange, 21* (1), 6-13.

Hanna, G. (1991). Mathematical proof. In D. Tall (Ed.), *Advanced mathematical thinking* (pp. 54-61). Netherlands: Kluwer Academic Publishers.

Harel, G. & Sowder, L. (1998). Students' proof schemes: Results from exploratory studies. *Research in College Mathematics Education.*

Lakatos, I. (1976). *Proofs and refutations: The logic of mathematical discovery.* Cambridge: Cambridge University Press,

Maher, C. & Martino, A. (1996). The development of the idea of mathematical proof: A 5-year case study. *Journal for Research in Mathematics Education, 27* (2), 194-214.

Martin, W. G. & Harel, G. (1989). Proof frames of preservice elementary teachers. *Journal for Research in Mathematics Education, 20* (1), 41-51.

National Council of Teachers of Mathematics. (2000). *Principles and standards for school mathematics..* Reston, VA: Author.

Peressini, A. (1999) "Confirming Mathematical Theories: An Ontologically Agnostic Stance," *Synthese*, vol. 118, 257-277.

Peressini, A. (2003) "Proof, Reliability, and Mathematical Knowledge", *Theoria* (3), part 3, 211-232.

Peressini, D., Borko, H., Romagnano, L., Yorker, C., Wooley, C., Hovermill, J., & Masarik, K. (2001). *The transition from student teacher to first-year teacher: A situative view of learning to teach secondary mathematics*. Symposium paper presented at the Annual Meeting of the American Educational Research Association, Seattle.

Peressini, D. & Knuth, E. (1998). Why are you talking when you could be listening? The role of discourse in the professional development of mathematics teachers. *Teaching and Teacher Education, 14* (1), 107-125.

Putnam, H. (1971) "Philosophy of Logic", in Putnam (1979), pp. 323-357.

Putnam, H. (1979) *Mathematics Matter and Method: Philosophical Papers Vol. I,* second edition (Cambridge: Cambridge U. Press).

Putnam, H. (1979a) "What is Mathematical Truth?", in Putnam (1979), pp. 60-78.

Quine, W.V.O. (1960) *Word and Object*, (Cambridge: M.I.T.).

Quine, W.V.O. (1969) *Ontological Relativity and Other Essays*, (New York: Columbia U. Press).

Quine, W.V.O. (1969a) "Epistemology Naturalized", in Quine (1969), pp. 69-90.

Quine, W.V.O. (1969b) "Existence and Quantification", in Quine (1969), pp. 91-113.

Quine, W.V.O. (1981) *Theories and Things* (Cambridge, Mass.: Harvard U. Press.)

Quine, W.V.O. (1981a) "Five Milestones of Empiricism", in Quine (1981), pp. 67-72.

Quine, W.V.O. (1981b) "Success and Limits of Mathematization", in Quine (1981), pp. 148-155.

Rav, Y. (1999) "Why Do We Prove Theorems?", *Philosophia Mathematica* 7:1, pp. 5-41.

Richards, J. (1991). Mathematical discussions. In E. von Glasersfeld (Ed.), *Radical constructivism in mathematics education* (pp. 13-51). Netherlands: Kluwer Academic Publishers.

Royden, H. L. (1988) *Real Analysis*, 3rd edition, (New York: Prentice Hall).

Rudin, W. (1986) *Real and Complex Analysis,* 3rd edition, (NY: McGraw-Hill).

Schoenfeld, A.H. (1998). Toward a theory of teaching-in-context. *Issues in Education: Contributions from Cognitive Psychology*, *4*(1), 1-94.

Schoenfeld, A. (1994). What do we know about mathematics curricula? *Journal of Mathematical Behavior, 13*(1), 55-80.

Simon, M. & Blume, G. (1996). Justification in the mathematics classroom: A study of prospective elementary teachers. *Journal of Mathematical Behavior, 15* (1), 3-31.

Steiner, M. (1978) "Mathematical Explanation", *Philosophical Studies* 34: 135-151.

Van Fraassen, B. (1980) *The Scientific Image* (Oxford: Clarendon Press).

Wheeler, D. (1990). Aspects of mathematical proof. *Interchange, 21*(1), 1-5.

Chapter 11

MATHEMATICAL PRACTICES IN AND ACROSS SCHOOL CONTEXTS

Jill Adler
University of the Witwatersrand, Johannesburg, South Africa

Abstract: This paper describes some mathematical practices in and across school contexts in South Africa. Through this description, I challenge a decontextualised notion of schooling and so too a decontextualised notion of mathematical practices in school. Through examples of mathematical practices within and across school contexts in South Africa, I illustrate what is well known in the field of sociology of education – that school mathematics and its associated practices are social at their core. They are a function of how knowledge in society is selected, transmitted and evaluated through schooling. These are socio-political processes and so vary considerably within and across social contexts, and inevitable function of the distribution of power and social control in society (Bernstein, 1996).

Key words: School mathematics, mathematical practices, South Africa

1. INTRODUCTION

This paper was conceptualized in response to an invitation to offer an educational perspective on mathematical practices to a Philosophy of Mathematics conference. My task for the conference was to provide some insight into mathematics in schools with an emphasis on the 'pictures' or 'view' of mathematics that results from such an education. In other words, if someone leaves school and has had some mathematics training, what does he or she think about mathematics[1]?

In this paper I describe some mathematical practices in and across school contexts in South Africa. Through this description I challenge the implicit assumption that there is or can be "a" mathematical education in school, and

[1] This was the direct request from the Conference Chair: Jean Paul Van Bendegem..

B. Van Kerkhove and J.P. van Bendegem (eds.), Perspectives on Mathematical Practices, 191–213.

an undifferentiated set of pictures of mathematics emerging from this education. In other words, the description I provide challenges a decontextualised notion of schooling and so too a decontextualised notion of mathematical practices in school. Through examples of mathematical practices within and across school contexts in South Africa, I illustrate what is well known in the field of sociology of education - that school mathematics and its associated practices are social at their core. They are a function of how, in Basil Bernstein's terms, knowledge in society is selected, transmitted and evaluated through schooling. These are socio-political processes and so vary considerably within and across social contexts, an inevitable function of the distribution of power and social control in society (Bernstein, 1996).

I begin the paper with a brief introduction to mathematics education in South Africa at this juncture. This provides for some awareness of the multiple influences on school mathematical practices, and how these come to take shape across diverse schools in the post apartheid era. I then describe some mathematical practices drawn from four different, but recent, research projects. Only some of this research had as a question, students' 'pictures' of mathematics. Each, however, has produced data that provide illuminating 'windows' on to mathematical practices in some South African schools. Together these provide empirical substance for my argument that a decontextualised notion of mathematical practice makes no sense from the perspective of school mathematics, if at all. School mathematical practices are just that: practices dialectically produced by both mathematics and schooling.

2. MATHEMATICS EDUCATION IN POST-APARTHEID SOUTH AFRICA

It is easy to understand and empathise with a strong push to political, social and economic change in post-apartheid South Africa. Building a democracy out of a deeply unequal, brutal and racialised society is a mammoth, some would say impossible task. There is enormous pressure for rapid and tangible change in the distribution of social and economic goods, with serious implications for the role of public education in this process. One critical element here is reform of the curriculum, including mathematics.

New National Curriculum Statements for Grades 1 – 9 (DoE, 2001) and more recently Grades 10 – 12 (DoE, 2002) have been published. They are components of what is called Curriculum 2005 (C2005). A long process of system-wide implementation is currently underway. In a detailed and insightful analysis of C2005, Graven (2002) describes four dimensions of mathematics evident in the elaboration of knowledge, skills and values in the Mathematics Statement for Grades 1- 9. School learners are to become proficient in key components of mathematics (e.g. they need to be fluent in basic computations, able to calculate and measure with speed and accuracy, and to use well established algorithms appropriately and efficiently). At the same time, this narrow skills-based dimension of mathematical profiency needs to be complemented by, indeed underpinned by, a grasp of key mathematical processes and practices (e.g. appreciating what is entailed by, and being able to, reason mathematically, being able to conjecture and generalize, and work in a disciplined way from assumptions through to deductions). These mathematical proficiencies and practices are to be acquired in ways that enable learners to apply their mathematical knowledge in real-world problem-solving (i.e. engage in mathematical modeling at various levels), and in ways that enable them to participate as critical citizens and in changing society (i.e. be able to engage with the ways in which mathematics currently 'formats' so much of social and economic life (Skovsmose, 1994)).

This is a bold and ambitious vision - for a new mathematical literacy on the one hand, and the production of mathematical excellence on the other. School mathematics in South Africa must pave the way for new mathematicians to emerge at the same time as make available for all learners, a flexible and useful form of mathematical knowledge. This dual goal for equity and excellence (Adler, 2002) is not peculiar to South Africa. It is in sharp focus, for example, in curriculum reform in the USA (NCTM, 2000). However, the challenges in South Africa take on a specific relevance given the huge disparities within the country across race, class and language background; and the increasing global divide between the so-called developed and developing or underdeveloped countries. There are constant pressures on South Africa as a developing economy to position itself and compete in the global economy, which in turn places demands on the school system to produce school leavers who are internationally competitive. Such competitiveness includes technological expertise, and so a tension in South Africa. Some of our rural schools do not yet have sufficient classrooms, or electricity and water, let alone, technological tools to support learning. Where and how, then, are resources to be distributed?

The magnitude of this challenge becomes clear in the face of the results of the 1995 Third International Mathematics and Science Study (TIMSS), and the 1998 TIMSS Repeat. Notwithstanding some methodological flaws in the study, for example, practical difficulties with constituting a full sample in South Africa, and the fact that South African students were tested in English or Afrikaans which are additional (i.e. second or third) languages for the majority of learners, the scores obtained by South African learners were way below those of all other participating countries (Howie, 1998; Howie & Plomp, 2002). Like its political counterparts in the United States and the United Kingdom, these results were used to make damaging claims about our mathematics teaching corps, so undermining those now charged with implementing a bold new vision. Mathematics teachers, who have been described over and again as 'lacking' in some way, and most recently in lacking mathematical knowledge, have to 'improve results' nationally and internationally while embarking on the implementation of a new curriculum that entails the enactment in practice of four mathematical roles and identities. Teachers need to instill mathematical proficiency, model the process and practices of mathematicians and applied mathematicians, and at the same time hold a critical understanding of how mathematics formats society. They need to meet these mathematical goals while dealing on a day to day basis with social inequality, limited resources for teaching and learning, and within a set of discourses about teaching that position them as lacking.

I paint this demanding landscape not to produce a sense of gloom, but to situate the discussion that follows. It is only with some appreciation of the demands on teachers and teaching in a context of social, political and educational change, that the mathematical practices I will describe can be understood. More pertinently for this paper, it is precisely because of the diversity and complexity that is current South Africa, that practitioners, whatever their field, appreciate just how much context matters, or, in the framework of the conference that motivated this paper, appreciate that school mathematical practices are a function of mathematics in schooling. What South African school leavers think about mathematics will be a function of what they have learned, where they have learned, and how the mathematics has been presented.

3. SOME SCHOOL LEARNERS' IMAGES OF MATHEMATICS AND THEIR MATHEMATICS TEACHERS

During 1999, a colleague in mathematics teacher education in the KwaZulu-Natal province had teachers in an in-service programme elicit from their Grade 9 and 10 learners, various images of mathematics, their mathematics teacher, and what he or she does in class (Parker, 1999). Parker's purposes here were bound up with a teacher development programme, which included providing opportunities for teachers to elicit from, and then reflect on, their students' conceptions of their mathematics learning in school. Despite my rather different purposes, I draw liberally from her work in this programme and her related study, as it provides a startling view of student images of mathematics.

The Kwazulu-Natal province is on the South East coast of the country, and includes one of the countries large urban cities (Durban). Schools in such urban settings are relatively well resourced. However, a far larger part of the province is rural and very poor. In between these two regional extremes are what were apartheid designed 'townships'. Relative to their urban counterparts, township schools are poorly resourced. The teachers Parker was working with came from this latter grouping and what she described as "typically disadvantaged schools and communities".

Teachers in Parker's project asked their learners to draw a picture of their mathematics teacher and what he or she does in their mathematics classes, as well as to write a short paragraph in response to each of the following:

- I think that learning mathematics is all about …
- When I think about doing mathematics, I feel ….

The responses below have been selected from Grade 9 and 10 learners. They are typical of the responses generated in the classes of all the teachers involved.

For most of the students, the response to **'I think that learning Mathematics is all about …'** included some reference to mathematics providing "chances of jobs". Mathematics for these learners is the gateway to future employment. No response included a mathematical description of learning or referred in any way to mathematical ideas or concepts. However, their responses to how they feel about doing mathematics reflect what Parker described as a deep 'love-hate' relationship with mathematics. The following are a selection of typical responses to: **When I think about doing mathematics I feel ….**

> … so good and proud because many jobs needs this mathematics. I feel good also because when you want to calculate sums it is easy when you do mathematics[2].
>
> Very glad because it is a basic … which will take me anywhere I like in future. And it is my favourite subject of which I feel better about learning it.

[2] It is important that readers know that the learners in these schools are first language Zulu speakers. For many, access to English is only in school. Hence the sometimes peculiar use of the English language.

When I think about doing mathematics I feel very happy because I think mathematics is bread forever.

I feel happy and am very proud of myself because mathematics is overly important subject in our days.

I like Mathematics, but I fail to do practice. The thing that makes me feel unhappy is because I cannot enjoy it. The teacher tries to teach me, but the problem faces me.

I feel so happy about it because I like this subject. Although I do not know it but … I want to know it.

When I think about mathematics I feel unhappy because I like mathematics and I spend time at mathematics. But at Test or when I write a test I feel unhappy because I fail a test. From Std 7 *(Grade 9)* until now.

I feel frightened to answer because maybe I'll get the answer wrong.

The pictures these students drew of their mathematics teacher and what he or she did in the mathematics classroom reinforce the fear of, and failure in, mathematics as reflected above[3].

[3] I am indebted to Diane Parker for generously allowing me to reproduce these pictures here.

(Girl - Grade 9)

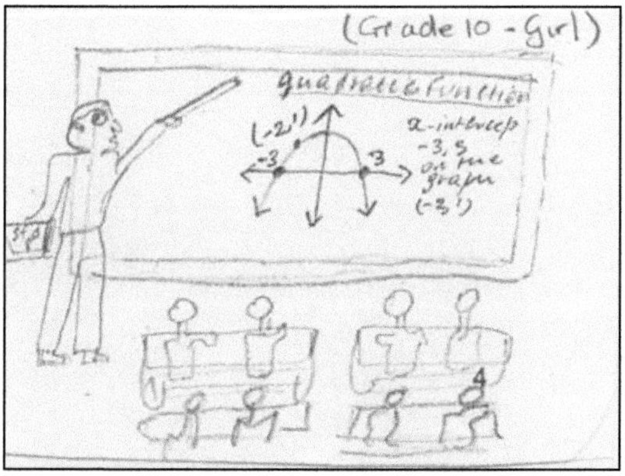
(Grade 10 - Girl)
(-2')
-3
3
(-3')
4

space below draw a pi
what he or she does in the classroom.
I WANT MY WORK
(Boy - Grade 10)
4

The six pictures above are distressing. There are interesting differences between the pictures drawn by the boys and girls. For the boys, maths teachers are male and heavy handed, wielding a stick literally and figuratively all the time. For the girls, stick wielding is part, but not all, of what their mathematics teacher does in their classroom. The mathematics classroom is a fearful and an unhappy place for some. But it is also a place where teachers take examples from the textbook and then explain them using the chalkboard. Students sit quietly and listen. And there are also teachers who help students, and students who work together.

What is fascinating here, yet so well known, is that similar images of fear and unhappiness are coupled with quite different experiences. And these emerge from similar sets of conditions. The students who drew these pictures were from similar school settings, and some will have been in the same classroom with the same teacher. They remind us that these are not depictions of classroom realities as such, but learners' experiences of such conditions and the practices within them.

The point for this paper is: what might constitute mathematical practices in these students classrooms if these are the images that school students produce in response to questions about mathematics and their mathematics teacher? The images drawn by the students above illustrate most vividly that no matter what mathematical knowledge might be taught and learnt, students are at the same time learning 'ways of being' in relation to mathematics (Boaler & Greeno, 2000). In these classrooms, mathematics, whether it is understood or not, is something to be revered and, at the same time, feared. When children learn mathematics, they don't only learn more or less "content" ... they also learn ways of being. Identities and subjectivities are produced.

The images may be described as extra-mathematical, illuminating students' emotional life in their mathematics classroom and what they think mathematics is for. Only a few indicate something about mathematics itself and its emergence in school practices. Their pictures include images of the teacher at the chalkboard pointing to inscriptions of mathematics, and so of mathematics as a knowledge domain that is accessed through show and tell.

What images emerge, then, when learners are specifically asked to produce some representation of their mathematical learning in school?

4. CONCEPT MAPPING AS A WINDOW INTO SECONDARY SCHOOL MATHEMATICAL PRACTICES

In a completely different study, Mwakapenda and Adler (2002) explored the extent to which some first-year university students were able to make connections among a selected number of key concepts in the South African secondary school mathematics curriculum. The study used concept mapping to investigate whether and how students were able to relate a selection of concepts and present these relations in a concept map. An underlying assumption in the study was that the kind of maps students would generate would reveal whether their experiences in school were of mathematics as a collection of fragmented bits of disassociated, 'inert' knowledge and skills, or whether the practices in which they were engaged had produced a more 'flexible' and connected form of mathematical knowledge (Boaler, 2002).

Students in three different mathematics courses were invited to participate in the study. There were a small number of volunteers from the Mathematics I Major class, and a larger group from an access programme for students who wished to enter the Faculty of Science but who did not have satisfactory Grade 12 mathematics scores. These latter students were studying Mathematics I over two years. In addition there were students in a Mathematics Foundation course offered to students in faculties other than

Science. All the students in the study had thus studied mathematics in school, but across wide-ranging school contexts and with very different levels of success.

In a paper that reports this research the authors discuss how most students included the concept of 'angle' in their maps, and related this to various other concepts presented to them. Below I include two quite

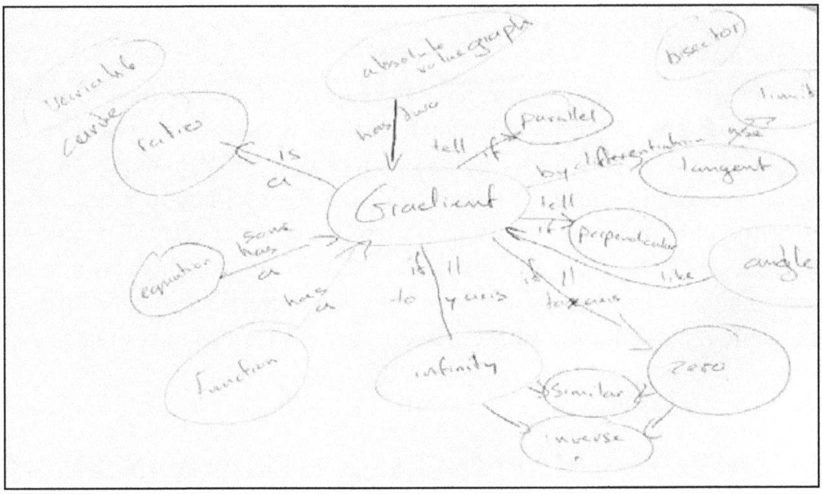

different maps, the first from a Mathematics Major student, the other from one of the students on the access programme. The first reflects a student who is able to take a central concept like 'gradient' and relate this, mathematically, to a wide range of other concepts, including 'angle'. This map reflects a connected knowledge of mathematics, and a flexibility in the student to link concepts in considered ways.

The second map below, and the explanation provided by the student in a follow-up interview illustrate what was the case for most of the students in Mwakapenda and Adler's study.These students produced maps with single links between concepts, rather than displays of complex networks of connected concepts.

Follow-up interviews with students provided opportunity for students to explain why they linked concepts in the way that they did. The interviews provided further insight into the ways in which students thought about mathematics. In the main, the descriptions given by students who produced maps similar to the second map above were more about the contexts in which they had learnt about various concepts, including angle, than the conceptual relationships between angle, say, and other concepts given to them. The access student who produced the second map explained the links he produced as follows:

[In] high school we are told that maths is divided into two, geometry and algebra. So we thought in geometry we can find angles. We can find angles. OK. Angles can be found in geometry. That's from maths anyway. And we thought of tangents. Tangents like the theorem in Grade 11 Theorem 9 talking about tangent. Yah. So we thought of, it was, they were both tangent and angles. So we thought if we could take this out of tangent to angles. Tangents can be found in angles.

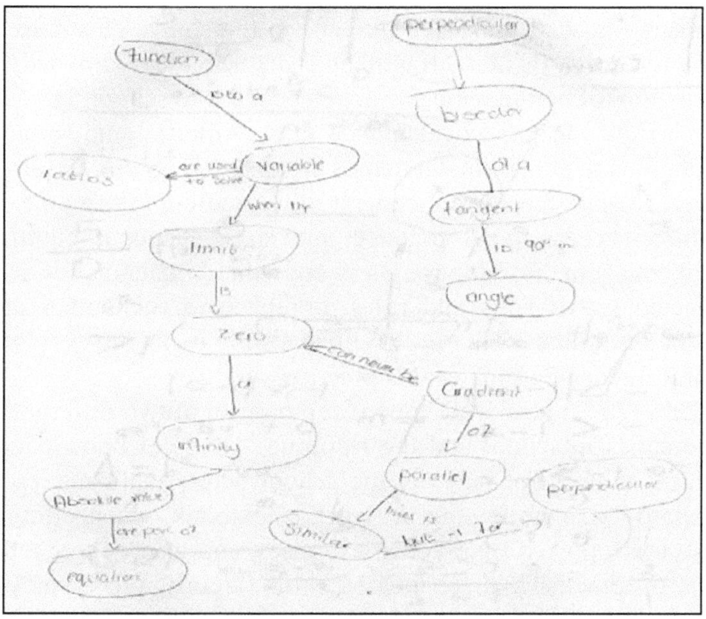

Upon being asked to provide an example to show what was meant by "tangents can be found in angles", he said:

It is not possible. We just thought that we can connect even though we did not know how to connect them. But they can connect… We thought of the Grade 11 theorem, Theorem 9. Yah, it was talking about angles and tangent. So we thought angles and tangent can be found in the same place so we just connected here. …

A number of other students made similar connections between angle and tangent e.g. "tangent have a angle… We sometimes say… like tan theta, sometimes you say tan ninety" and "perpendicular bisector of a tangent is 90 degrees in angle".

It seems that for these students', linkages between various mathematical concepts learned in secondary school were a function of remembering how you say things in the mathematics class. A different student explained the links she made by referring directly to their mathematics lessons in school.

> I don't remember the theorem… When we were doing maths at high school I mean the teachers keep on mentioning those words. … it was quite a long time back. But we know like in geometry, angles, tangent and parallel. If you talk about angles, it's just something we know in our mind. We just know that.

As mentioned, the second map is illustrative of the kinds of maps produced my most of the other students in the study. This map, and others like it, is reflective of inert, fragmented mathematical knowledge, and so a particular form of mathematical practice peculiar to schooling.

The insights from Mwakapenda and Adler's study into students' representations of school mathematics are distressing, but, unfortunately, not surprising. Despite the current context of curriculum reform in South Africa, the dominant pedagogical practice and so educational culture of many secondary mathematics classrooms is one where mathematics is taught as a disconnected set of facts and rules, unrelated to each other and to other knowledge disciplines. In Graven's analysis (Graven, 2002), mathematics as procedural fluency dominates this practice.

Boaler's (2002) now well-known comparison of two different approaches to mathematics in school and the resulting "forms of knowledge" acquired by learners is pertinent here. In the school where mathematics was taught "traditionally" i.e. following a typical textbook where definitions and examples are followed by practice exercises, and where there tends to be an emphasis on practising disconnected skills (algorithmic fluency), students' mathematical knowledge was inert and fragmented. This inert form of mathematical knowledge was contrasted with the kind of knowledge students acquired in a school where the approach to mathematics was considerably more open-ended and problem-oriented. In this school mathematics was engaged as a way of thinking, as well as a tool for problem-solving. Students in the school with the 'traditional' approach to learning mathematics were less able (than their peers in the latter school) to connect across mathematical areas and less able to use their mathematics in novel situations. Boaler's work provides a research-based explanation for diverse learners' experiences of mathematics, and an explanation that points to the critical role of teachers and teaching of mathematics, but locates the problem in the approach to mathematics rather than in the teacher per se.

In Mwakapenda and Adler's study the maps and links students were able to make are indicators of the way in which learning is deeply tied to the context in which it takes place. It is reflective of the findings in Boaler's far larger and more extensive study and so of her argument that substantively different pedagogical approaches to mathematics will result in learners emerging from their learning experience with different forms of knowledge. In her more recent research, Boaler has extended this analysis to notions of 'identity' and the argument that students emerge not only with different forms of knowledge, but as significantly, with different ways of being mathematical, different mathematical identities (Boaler & Greeno, 2000).

Hence further support for the argument in this paper. There is no singular set of school mathematical practices. When learners leave school they emerge with forms of knowledge related to pedagogical practices in their school. The mathematics they learn in school is not a collection of mathematical topics or content, but rather a set of social practices. School mathematical practices vary, and with these, then, come very different images or pictures of mathematics, and so too very different ways of being mathematical.

5. SNAPSHOTS FROM CLASSROOM PRACTICES WHERE TEACHING IS DIRECTED TOWARDS MORE FLEXIBLE MATHEMATICAL KNOWLEDGE

With the images of mathematics and its teaching and learning in schools in South Africa as portrayed above, and in the context of curriculum change and huge pressures for tangible transformation, a critical challenge for mathematics education in South Africa is how to shift dominant practices that produce at once reverence and fear of mathematics, fragmented and inert forms of knowledge, and mathematical identities that disassociate mathematics and meaning. As mentioned earlier, since the mid-1980s, multiple projects emerged in the field of mathematics teacher education across the country. In various ways all worked in support of changing classroom practices and thus changing mathematical practices and so too images and pictures of mathematics that learners take away with them when they leave school. Most of these projects have become more established programmes since 1994 – no longer are they outside of state reform process, but instead part and parcel of a national agenda to support the development of more flexible forms of mathematical knowledge in school learners.

In the remainder of this paper I will draw on images of mathematics classroom teaching that illustrate shifting approaches to mathematical knowledge and to pedagogy. These highlight different possibilities for students' views of mathematics than those portrayed above. I will start with two snapshots from a research project related to a professional development programme for mathematics teachers in primary and secondary, rural and urban schools in two different provinces in South Africa (Adler & Reed, 2002). This study followed 10 mathematics teachers over three years while they were participants in the programme and after they had exited from the programme. The research reports on teachers' 'take-up' from the programme, in an attempt to describe what it was teachers' selected and attended to in their learning (as opposed to whether and how they had changed their practices in particular ways). The authors are critical of teacher education research that anticipates uniform learning across teachers, and so inevitably constructs a discourse of teachers failing to 'reform' their practice. Adler and Reed demonstrate instead, just how diverse take-up can be. And in this diversity, at the same time, there were also some similar aspects to teachers' take-up that illuminate school mathematical practices in a context of reform.

Remembering the images drawn by the Kwazulu-Natal students in Parker's study, and the inert knowledge portrayed in students' concept maps, snapshots of mathematics classrooms[4] in the study reported in Adler and Reed suggest a different mathematics classroom experience.

In these classrooms, students are engaged in doing and communicating mathematics as well as listening to the teacher. The teacher is observant of what learners are doing, and identifies areas that require whole class mediation (e.g. the meaning of the power of 0). It is likely that the students

[4] These are freeze frames from videotapes of a range of classrooms and lessons.

in these classrooms, if asked to do so, will draw very different images of their mathematics classrooms and their teachers.

There is much else of value in the wider study about teachers' learning in a context of curriculum change, all of which falls beyond the scope of this paper. Indeed, one part of the wider study has been reported by Brodie, Lelliott and Davis (2002) and reveals the difficulties involved in substantive change in school mathematical practice.

I continue instead with an episode in a Grade 8 classroom of a teacher, Sue, who over time had developed a sophisticated activity-based mathematical practice[5]. The mathematical practices in her classroom are productive of students who think and talk about mathematics in ways quite different from those reflected so far in this paper. At once, this classroom supports my argument for a contextualised notion of mathematical practice, and a heartening insight into possibilities for students emerging with ways of being mathematical that are currently highly valued in the curriculum reform process sweeping classrooms and educational systems in many countries.

6. A VIEW OF 'BEING MATHEMATICAL' IN SCHOOL MATHEMATICS CLASSROOMS

The research from which the episode in Sue's class is drawn was conducted in the early 1990s and before the demise of the apartheid state in South Africa. As discussed earlier, attempts at shifting mathematics curriculum practices in South African schools date back to the 1980s. Sue was teaching in a private school, well-known for its progressive stance both pedagogically and politically. The school was well-resourced and, as a whole, supported activity-based and inquiry-based learning.

As part of a sequence of tasks related to angles and triangles, Sue gave the activity in the box below to her Grade 8 classes.

In one of these classes where there were 16 students, Sue had them work on this task in pairs. They were required to record their responses and be prepared to present these and their justifications for them to the rest of the class. I am going to focus my discussion here on the student responses to the highlighted task – the possibility of a triangle with two obtuse angles.

[5] Sue participated in a differently focused research project, one concerned with mathematics teachers' knowledge of the practices in multilingual classrooms. For more detailed discussion of Sue's teaching and her knowledge of this, see Adler (1997) and Adler (2001).

If any of these is impossible, explain why, otherwise draw it.
• Draw a triangle with 3 acute angles. • Draw a triangle with 1 obtuse angle. • ***Draw a triangle with 2 obtuse angles.*** • Draw a triangle with 1 reflex angle. • Draw a triangle with 1 right angle.

First, however, it is useful to reflect on both the ***mathematical and pedagogical intentions*** possible to read into this task. The ***mathematical content*** demands of the task are most transparent: learners are asked to recall different types of angles and then use these to explore and consolidate their understanding of the properties of various types of triangles. More obscure, but clearly embedded in the task is the building of ***mathematical proof***. Learners are asked to explain why some of the cases above are 'impossible' i.e. their explanation needs to convey why it can *never* be the case that ... The response here needs to be such that it holds in all cases, and not be a specific example. In addition, the task creates the expectation that students will *reason, justify and argue* as they develop their responses; that they will *communicate* their mathematical ideas verbally, orally and in writing, as well as pictorially. There is thus the added expectation that the learners can and will *represent* their ideas in different ways.

A range of mathematical practices is embedded in this task and so too the production in students' of images of mathematics that differ substantially from those illustrated in the earlier parts of this paper. The construction of the task is such that students are given a reason to reason mathematically (Ball and Bass, 2001). Here we have an illuminating example of a practice where school learners are confronted with a 'big idea' in mathematics (e.g. proof). This task makes high-level mathematical demands on learners and as such is a good exemplar of the reform goals for promoting 'excellence' in school mathematics (e.g. NCTM 2000).

Elsewhere (Adler, 1997; 2001), I have described in detail Sue's ***pedagogical intentions*** in this lesson and her teaching in general (based on the empirical data in the research project of which Sue's lessons were part). Alongside her concern for promoting excellence in her students' mathematical thinking and working, Sue embraced the wider educational concerns of equity and student voice. She structured tasks and classroom interaction so that all students could be actively engaged in their learning, and encouraged all to participate fully in class. This is apparent in her having students work on the task in pairs, and then requiring that they communicate their thinking to the rest of their classmates. In the whole-class discussion, students were asked to comment on the responses offered by their classmates. Sue's pedagogical practice, as briefly described here, is reflective of the dual demands contained in the reform process in South Africa and elsewhere, for excellence and equity. High-level mathematical demands are made in class, and in ways that these are accessible to all learners.

The students' responses, and what Sue does with these within the time constraints of a school lesson, is illustrative of the challenges entailed in such a vision for mathematics in school, and so too of possibilities for school mathematical practices.

Across the pairs in one of the classes, Sue's students produced the following responses to the 'two obtuse angles' task:

- Some said: It is impossible to draw a triangle with two obtuse angles, because you will get a quadrilateral. And they drew:

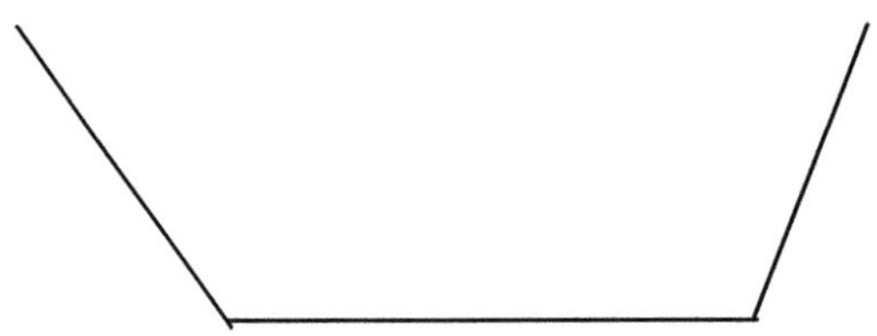

- Others reasoned as follows: an obtuse angle is more than 90 degrees and so two obtuse angles gives you more than 180 degrees, and so you won't have a triangle because is angles must add up to 180 degrees
- Joe and his partner reasoned in this way: If you start with an angle say of 89 degrees, and you "stretch it", the other angles will shrink and so you won't be able to get another obtuse angle. They drew:

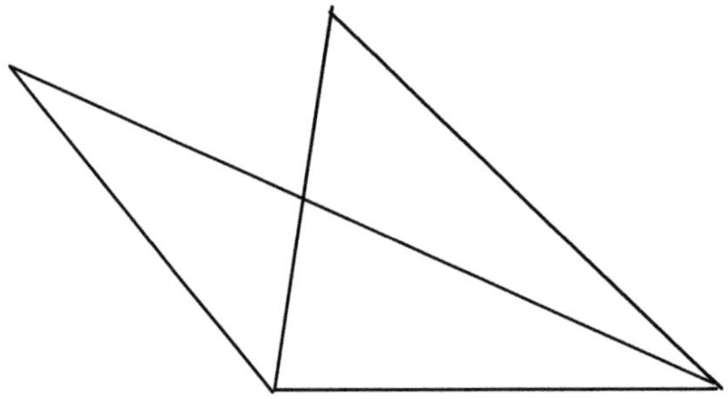

As students worked on their responses in their pairs, Sue moved across the classroom, asking each pair questions like: Explain to me what you have drawn/written here? Are you sure? Will this always be the case? And she asked Joe in particular whether his explanation of stretching one of the angles would hold if he started with a triangle with one obtuse angle and then stretched one of the other angles. She was not convinced of the generality of his response.

In the last quarter of the lesson, individual students volunteered to share their solutions with the rest of the class, leading to the public display of the above three different responses. During this whole-class public discussion, Sue asked questions of the class to ensure all followed each of the explanations, and provided supportive and encouraging comments on the diverse possible solutions presented. In terms of her mathematical and pedagogical intentions, students here had multiple opportunities to consolidate their understanding of types of angles and triangles, as well as to reason, argue and communicate mathematically. Student voice was apparent through the lesson, with visible encouragement of their mathematical thinking in its diverse forms.

The 'big idea' of proof however, was backgrounded in the public whole-class discussion. There are many possible reasons for this, ranging from the practical - constraints of time within a single lesson, to the mathematical – the mathematical demands on the teaching and development of notions of proof and proving (providing an answer that 'holds in all cases'). Here it is insufficient that the teacher has a deep grasp of the notion of proof. What is required here is that this notion is managed in this class, where the communal mathematical knowledge with which the class can work to prove that there cannot be two obtuse angles in a triangle can probably only rest on the prior establishment of the sum of the angles of a triangle being 180 degrees. The mathematical entailments of accepting the visual representation of two connected obtuse angles as a proof let alone Joe's dynamic explanation are unclear.

What Sue actually did, however, is not the main focus of this paper. The critical point for describing in some detail student productions in this class is that they are indicative of a classroom environment where mathematical reasoning is encouraged. The images of mathematics and teaching that Sue's students might draw are likely to be quite different from those depicted earlier in the paper. If asked to produce concept maps, they are likely to produce maps built on connections between concepts.

The discussion of what Sue did do, and the mathematical and pedagogical demands on her teaching that follow from the kind of task she set her learners, is a window into school mathematical practices valued in reform initiatives in South Africa and elsewhere.

7. WHAT DO THESE VARIED SNAPSHOTS AND DESCRIPTIONS ADD UP TO?

In this paper I have selected to present a range of images, descriptions and snapshots of mathematics, mathematics teachers and teaching as produced by students and within classrooms across a range of classroom contexts in South Africa. My purpose in doing this was to reveal what is well known: that schooling practices are inherently social, and so too mathematics within schooling. There is no singular image of mathematics and mathematics teaching than will emerge across ranging classroom contexts, or in Boaler and Greeno's (op cit) terms, no single way of being mathematical that can emerge from such ranging practices. School mathematics practices are as diverse as the socio-cultural, political and economic contexts in which they function.

It is the field of the sociology of education that has described and explained the social nature of school practices and the consequences for learning. I draw briefly here from two interesting texts in this field so as to foreground the theoretical orientation that has informed the discussion in this paper. The first is the influential work of Basil Bernstein (1996) whose description of pedagogic discourse provides the tools with which to understand how it is that disciplinary knowledge is transformed (recontextualised in his terms) by the demands of pedagogic practice. Bernstein explains how in any pedagogic practice there are always two different but interweaving discourses in play: instructional discourse, which is turned towards disciplinary knowledge; and regulative discourse, which is turned towards the demands of teaching in school. He has argued theoretically and shown empirically how, because of the public demands of schooling, and the location of schooling in the presiding socio-political order in any society, regulative discourse is overdetermining of the nature of pedagogic discourse. It subordinates instructional discourse to its moral and public obligations.

As regards the diverse mathematical images or image inferences discussed in this paper, Bernstein's description of pedagogic discourse suggests that the way mathematics comes into being in any classroom is a function as much, if not more, of the educational code in operation as it is of the nature of mathematics. Schooling in and across diverse contexts will lead to diverse social practices and so too the images of mathematics and mathematics teaching with which learners leave school.

Moving from a within country perspective, the notion of the social nature of schooling is profoundly captured by Robin Alexander in his book "Culture and Pedagogy" (Alexander, 2000). Alexander's book is a detailed comparison of elementary teaching and learning (pedagogy) across five different countries: The UK, USA, Russia, France and India. He is able to show with rigour, depth and breadth just how pedagogy and culture intertwine in the production of school teaching and learning practices. His study is thoroughly convincing of the argument that any analysis of pedagogy, and so of teaching and learning in school, can only be fully understood in relation to the cultural formations within which this is inserted.

Both Bernstein and Alexander help explain why it is that a decontextualised notion of mathematical practices makes little sense in the context of school mathematics. Images of mathematics that school learners take with them will, inevitably, be diverse and a reflection of the culture and context in which they are learned.

8. POST SCRIPT: IMPLICATIONS FOR MATHEMATICS EDUCATION

I would like to conclude this paper, as I did the presentation, with some discussion on the implications of a conception of diverse mathematical practices in school in a context of mathematics education reform that is sweeping across many countries at present.

In South Africa, mathematics teachers at all levels, and particularly Grades 1 to 9 are working on their teaching in the interests of new curriculum requirements. They face, on a daily basis, the two compelling goals of excellence and equity. And many will attest to struggling to bring to fruition the kind of mathematical reasoning desired. At the same time, national and provincial department officials, together with the policy workers in education in the country, observe what they see as inadequacies in teachers' mathematical backgrounds. A discourse of 'a dumbing down of the curriculum' on the one hand, and the mathematical inadequacies of teachers on the other permeates policy and political discourses in South Africa, and is similarly present in the way in which many mathematicians understand the educational changes underway.

The point I wish to assert here is that mathematical demands of 'excellence and equity', of high-level mathematical thinking for all, and a curriculum that values 'big ideas' in mathematics (such as is reflected in the single task set by Sue) are not the typical fare of teacher preparation and development programmes. Nor are they the focus of attention in pure mathematics courses in undergraduate study of mathematics. There is much to do to work towards the kinds of mathematical practices now valued for school learning, much in teachers' work that reflects that their knowing of mathematics is and needs to be of a special kind, a kind attuned to the needs of teaching and learning within the constraints of schooling.

The question of the mathematical work of teaching, of taking the notion of mathematical practices, and understanding better what these are in diverse school classrooms is a task for all concerned with mathematics: philosophers and educationists alike.

ACKNOWLEDGMENTS

A presentation form of this paper was presented at the International Conference on the Philosophy of Mathematics, on Perspectices on Mathematical Practices, *24 - 26 October, 2002, Centre for Logic and Philosophy of Science, Brussels University (VUB), Belgium*

REFERENCES

Adler, J. (1997). A participatory-inquiry approach and the mediation of mathematical knowledge in a multilingual classroom. *Educational Studies in Mathematics, 33,* 235–258.

Adler, J. (2001). *Teaching mathematics in multilingual classrooms.* Dordrecht: Kluwer.

Adler, J. (2002) Global and local challenges of teacher development. In Adler, J., & Reed, Y. (Eds.) (2002). *Challenges of teacher development: An investigation of take-up in South Africa.* Pretoria: Van Schaik. Pp. 1-16.

Adler, J., & Reed, Y. (Eds.) (2002). *Challenges of teacher development: An investigation of take-up in South Africa.* Pretoria: Van Schaik.

Alexander, R. (2000) Culture and Pedagogy. Oxford: Blackwells.

Bernstein, B. (1996). *Pedagogy, symbolic control and identity: Theory, research and critique.* London: Taylor & Francis.

Ball, D. and Bass, H. (2000) Interweaving content and pedagogy in teaching and learning to teach: Knowing and using mathematics. In Boaler, J. (Ed.) *Multiple perspectives on mathematics teaching and learning.* Westport: Ablex Publishing. Pp.83-104.

Boaler, J. and Greeno, J. (2000) Identity, agency and knowing in mathematics worlds. In Boaler, J. (Ed.) *Multiple perspectives on mathematics teaching and learning.* Westport: Ablex Publishing. Pp.171-200.

Boaler, J. (2002) *Experiencing school mathematics.* Mahwah: Lawrence Erlbaum.

Brodie, Lelliott and Davis 2002 Brodie, K., Lelliott, A., & Davis, H. (2002) Forms and substance in learner-centred teaching: teachers' take-up from an in-service programme in South Africa. *Teaching and Teacher Education, 18*(5), 541-559.

Department of Education (DoE) (2001). *Revised National Curriculum Statement: Overview.* Pretoria: Department of Education.

Department of Education (DoE) (2002). *National Curriculum Statement: FET Mathematics* . Pretoria: Department of Education.

Graven, M. (2002) *Mathematics teacher learning: Communities of practices and the centrality of confidence.* Unpublished doctoral thesis. Johannesburg: University of the Witwatersrand.

Howie, S. (1998). TIMSS in South Africa: The value of international comparative studies for a developing country. In J. Adler (Ed.), *Perspectives on the Third International Mathematics and Science Study* (Proceedings of a National Seminar, Mathematics Education Development Programme, pp. 22–40). Johannesburg: University of the Witwatersrand.

Howie, S and Plomp, T. (2002) School and classroom level factors and pupils' achievement in mathematics in South Africa: A closer look at the South African TIMSS-R data. In Malcolm, C and Lubisi, C. (Eds.) *Proceedings of the 10th Annual conference of the Southern African Association for Research in Mathematics, Science and Technology Education*. Durban: University of Natal. Pp. 116-123.

Mwakapenda, W. and Adler, J. (2002) "Do I still remember?": Using concept mapping to explore student understanding of key concepts in secondary mathematics. In Malcolm, C and Lubisi, C. (Eds.) *Proceedings of the 10th Annual conference of the Southern African Association for Research in Mathematics, Science and Technology Education*. Durban: University of Natal. Pp. 60 – 67.

National Council for Teachers of Mathematics (NCTM) (2000). *Principles and standards for school mathematics*. Virginia: NCTM.

Parker, D. (1999) *Images of mathematics and mathematics teachers.* Unpublished paper presented at a one-day conference on Researching formalised INSET in mathematics, science and English at the University of the Witwatersrand. Johannesburg.

Skovsmose, O. (1994) *Towards a philosophy of critical mathematics education*. Dordrecht: Kluwer.

Chapter 12

THE IMPORTANCE OF A JOURNAL FOR MATHEMATICS TEACHERS

Ad Meskens
Hogeschool Antwerpen, Departement BLS, Lerarenopleiding

Abstract: The author gives a personal view on the importance of a mathematics journal as an indispensable intermediary between mathematics teachers and the research community.

Key words: Journal, education, mathematics.

1. INTRODUCTION

This article is based on the experiences and ideas held by the author, and does not claim any scientific validity in the strict sense. It does have the intention though of contributing to – or perhaps starting – a much needed debate on the role of mathematics teachers in the mathematical community at large. The basic claim is that mathematics teachers are an underestimated force and offer a vast potential for mathematical research, albeit at a "low" level.

From 1989 to 1994, the author has been the editor-in-chief of *Wiskunde en Onderwijs* (Mathematics and Education), the journal of the Flemish Association of Mathematics Teachers (FAMT), whose policy has always been to be a journal *for* teachers, *by* teachers. This is of course a very broad mission statement, but it seems that journals of this kind all over the world honour more or less a similar policy. What then does this mission mean in terms of editorial practice? As far as I can see, it translates into three major categories of contributions:

1. articles by teachers, relating to class room experiences or presenting material they have themselves developed;

B. Van Kerkhove and J.P. van Bendegem (eds.), Perspectives on Mathematical Practices, 215–230.

2. articles by professional scientists, introducing new research that is either potentially relevant for use in the classroom, or serves as valuable background information;
3. contributions to the problem corner and articles which are, in one way or another, a response to some of the questions already posed there.

In sections 3 to 5, we shall have a closer look at a selection of material which has been published in the journal, holding on to this categorization. Although *Wiskunde en Onderwijs* is taken as a source, readers of the present paper will probably recognize the (type of) material from their own teacher's journal. Perhaps short of a physics journal, the third type of article in particular is very specific to a mathematics journal. First, however, a couple of general remarks on the journal.

2. SOME GENERALITIES

Wiskunde en Onderwijs is a journal with a circulation of about 1000 copies. It once flirted with a circulation of 2000, in the late nineteen seventies, without ever actually reaching this number. One can think of two reasons for this high circulation back then. The first is that, following a neo-Keynesian "deficit spending" policy, the then government largely absorbed unemployment of university and polytechnics graduates in secondary education, which resulted in a higher than normal number of teachers. Second, and for our topic more importantly, we were in the heyday of the "modern mathematics", i.e. after the introduction of set theory and related concepts into the class rooms. A great many teachers were not accustomed with these concepts, as they had had no formal training in set theory. Therefore, there was an urgent need for information on the subject, particularly on its didactics. *Wiskunde en Onderwijs* was the obvious forum to cater for such needs[1].

As in the nineteen eighties the Belgian government hailed a more "Thatcherian" overall policy, unemployment among teachers rose. At the same time, banks and insurance companies were actively engaging mathematicians. Consequently, the number of mathematicians choosing for an educational career dropped by the year, and so did circulation of *Wiskunde en Onderwijs*. And anyway, the generation which had not been

[1] Apparently, 'revolutionary' changes in the way the subject mathematics is taught generate discussion. Indeed the 'modern' ways of teaching Euclid and geometry was one of the reasons why The Mathematical Association was founded in the late 19th century. See M.H. Price, *Mathematics for the Multitude*, The Mathematical Association, Leicester, 1994.

acquainted with set theory was reaching the age of retirement. Surprisingly, the – ongoing – computer and IT revolution, started in the late eighties, did not succeed in curbing this evolution, although quite some attention was devoted to it. One might perhaps wonder whether the journal does still reach its intended audience, or whether there are possibly alternative or complementary reasons, such as a general oversaturation with information, particularly via the web.

Clearly, a journal for mathematics teachers, especially one published for a fairly little political region (Flanders) and written in a "small" language (Dutch), needs to be aware of the public it caters for. In our particuler context, there is very little to no room for different journals catering for the same specific subgroups. Therefore, *Wiskunde en Onderwijs* tries to cater for all mathematics teachers in secondary and higher education. Table 1 gives an impression of what this means for the different types of secondary school. Again, this is the situation specifically for Flanders, although it may well be recognized in broad terms by mathematics teachers around the globe.

	General	Technical	Vocational
Grade 5-6 (age 16-18)	University (MA or MSc)	University (MA or MSc)	Polytechnic (BA or BSc)
Grade 3-4 (age 14-16)	University (MA or MSc) OR Polytechnic (VocBA or VocBSc)	University (MA or MSc) OR Polytechnic (VocBA or VocBSc)	Polytechnic (VocBA or VocBSc)
Grade 1-2 (age 12-14)	Polytechnic (VocBA or VocBSc)	Polytechnic (VocBA or VocBSc)	

Table 1

University = University trained teacher (4 years scientific training + 1 year teacher training)
Polytechnic = Polytechnic trained teacher (3 years teacher training)
Between brackets are the new designations applicable after the Bologna Declaration reforms.
VocBA (Sc) = Vocational Bachelor of Arts (Science).

Moreover, there is to be noted a strong differentiation in the last two years (grade 5-6), with pupils being allowed to choose between maths courses of 2, 4, 6, up to 8 hours a week. Obviously, the way mathematics is taught in the latter course is different from that in the first one. Indeed even the topics may be different[2]. From this outline it can easily be inferred that the audience of *Wiskunde en Onderwijs* is a very fragmented one.

Before taking a closer look at some of the articles in the three categories mentioned above, we should first address another question, *viz.* who does the writing? Although *Wiskunde en Onderwijs* is a journal *for* teachers *by* teachers, it is an established fact that there is a hard core of regular authors. Indeed a survey of articles published over the last 25 years shows that there are hardly any authors who have published just once. On the other hand, the editors of the journal – and the present author is no exception to this – always complain that there is not enough material to be published, and are continuously looking out for new authors. So how may this seemingly paradoxical situation be explained? In general, the editorial board has found it very difficult to persuade people to contribute. Although officially, the overall sentiment seems to be one of being "not clever enough" or the like, more often than not, hidden behind this lies the author's fear of becoming the laughing stock of his colleagues. A mostly misplaced sentiment, for the editorial board meticulously guards over the quality of the papers. What we see, though, is that once someone has crossed the Rubicon of his fears, (s)he gets a feeling of wanting to do this more often, and that may indeed be the start of a number of articles. However, once the author gets known, so do his or her views. Unwittingly, an author begins to repeat him/herself, and then the editor thinks it is time for new ideas, new authors,... A final remark concerning authors is that it is striking how many of them are upwardly mobile, in the sense that especially those who are regular contributors tend to become headmasters, inspectors, members of the attainment target board, lecturers at pedagogical colleges, etc. This could be due to the fact that contributing to the journal would mostly imply a critical stance towards the existing 'system', while the author might at a point begin to realize that changes can best (or even only) be done from within.

3. ARTICLES RELATING TO CLASS ROOM EXPERIENCES OR MATERIAL

Frank Laforce, the first president of the FAMT, and one of the prolific writers in *Wiskunde en Onderwijs*, once came back from holidays with the idea of using his snapshots in the math classroom. He succeeded remarkably

[2] National curricula are very different, yet mathematical curricula obviously have a common core: algebra, calculus and trigonometry. See G. Howson, *National Curricula in Mathematics*, The Mathematical Association, Leicester, 1991.

well. Look at the drawing of a Romanesque church somewhere in France (Figure 1). Notice that the base of the tower is a square, while the tower is an octogon. How do you construct a regular octogon inscribed in a square? It is the beginning of a small but beautiful article on regular n-gons and n-heders. This article falls in the subcategory of applying mathematics to things we see around us.

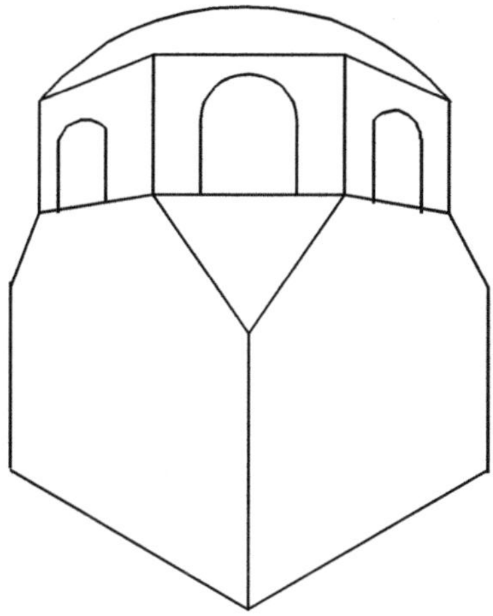

Figure 1

Taking this one step further, authors go looking for specific subjects that may illustrate or enliven a mathematical concept, which constitutes another subcategory. Figures 2 and 3 give nice illustrations of this. We see possible applications of trigonometric formulae in right angled triangles (Figure 2), and a problem in spatial reasoning, the question being what tourist took which picture at Gizeh (Figure 3). These examples were taken from an article by J.P. Daems on how to interest pupils in vocational classes[3].

[3] J.P. Daems, "Ideeën voor het beroepsvoorbereidend jaar", *Wiskunde en Onderwijs* **14**:120-42, 1988.

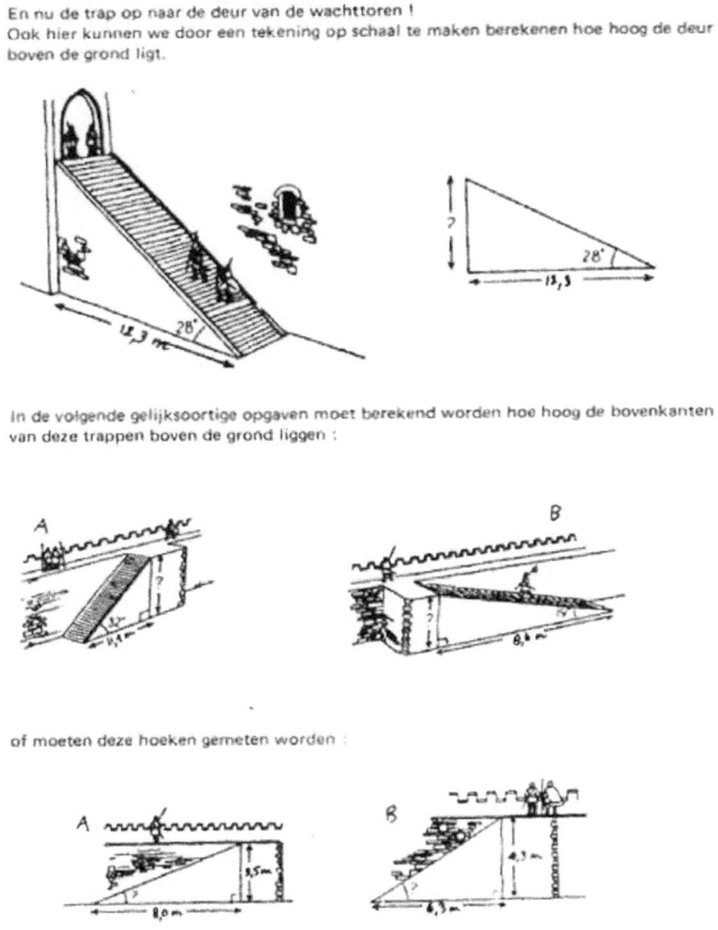

Figure 2

Sometimes this procedure is taken a step too far, and instead of giving a problem which is termed realistic or which can readily be seen as a simplification, problems become fairy tales. Usually, these examples not only are uninteresting to the pupils but they alienate them from the subject even further. Take for instance the problem: "In the Wild West cowboys were given a rope of given length and four poles to demarcate their plot of land. What is the plot of land with the largest area?" The example is real enough, as it was proposed by an author. Any editor will now face the difficult question of whether to publish or not. Given the restraints mentioned above, he may give the author a fair chance. Unfortunately, these articles get very few rebuttals. If there is some kind of rebuttal, it hardly generates any responses. Apparently argumentation is not really part of our tradition[4].

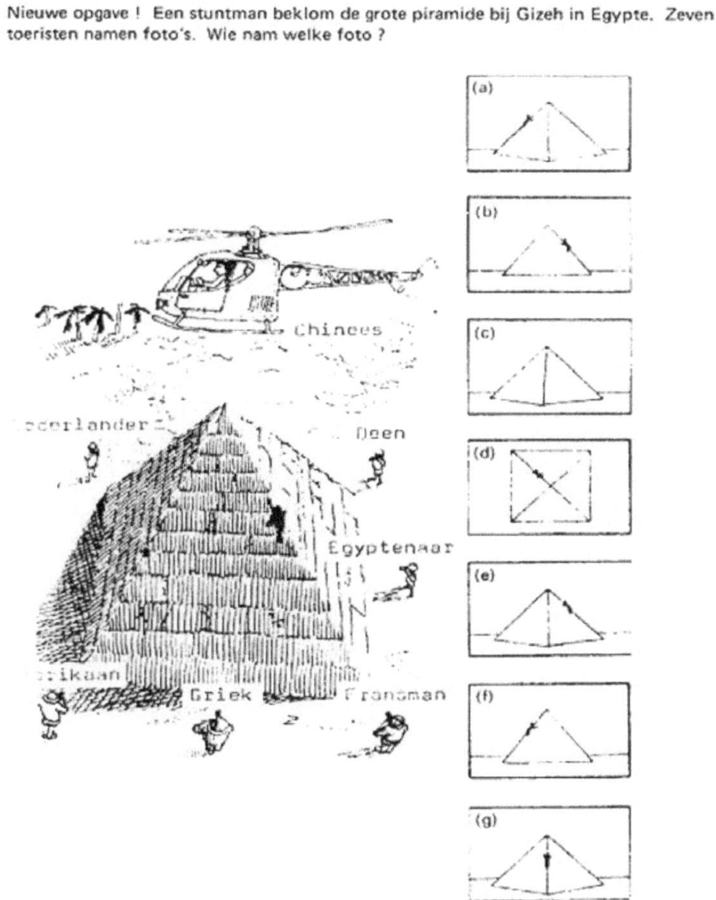

Figure 3

The cowboy example is seen by pupils as a story designed around some piece of mathematics, and not too cleverly for that matter. Yet a similar story *can* give credence to lessons, if it is told in another guise, that of an ancient myth for instance. In the story of queen Dido, she is given a plot of land that can be encircled by a cow's skin. It is an example of an isoperimetric problem just as the one of the cowboys. The story hinges on the cunning of

4 H. Van Maldeghem, "Enkele aspecten van het meetkunde-onderwijs", *Wiskunde en Onderwijs* **24**: 9-17, 1998.

queen Dido, which immediately gives it another flavour, and also establishes some links with other subjects, such as Latin, Greek and literature.

Another subcategory is the one in which an author does not write what would be properly called an article, but simply submits the course he has prepared for his classes. These texts may need some introduction written by the editor, but apart from the fact that one should of course not expect any "new" mathematical insights, they are usually very good indeed. These are texts aimed at young pupils, and they draw together many exercises which the author has encountered during his career, presented in a new fashion. Some authors are surprisingly good at this. E.g., *Wiskunde en Onderwijs* received numerous entries of this type written by Toon Huybrechts, his article "Fie" [phi] having been his *piece de résistance*[5]. This was in fact the first article for which various other teachers explicitly asked permission to copy and to use in class. Mostly, this formality is "forgotten", and the material is used anyway.

A further subcategory is that in which new ways of introducing theoretical concepts are shown. Often these are disguised in a theoretical framework. The succesive theorems are put together in such a way that the introduction of the concept becomes clearer. In figure 4, e.g., we find a very theoretical way of demonstrating a generalization of systems of linear equations[6]. On the left hand side we find an interpretation of straight lines as affine hyperplanes. Here the straight lines are considered to be (n-1)-dimensional affine subspaces given by the system

$$\begin{cases} L \Longleftrightarrow ax + by = c \\ Q \Longleftrightarrow a'x + b'y = c' \end{cases}$$

with an equivalent matrix expressed as

$$\left(\begin{array}{cc|c} a & b & c \\ a' & b' & c' \end{array}\right) = \left[M \,|\, B \right].$$

In the right hand side column the equations are considered as a one-dimensional affine subspace of the plane, which is algebraically translated into a pair of parametric equations.

[5] T. Huybrechts, "Fie", *Wiskunde en Onderwijs* **17**: 3-108, 1991.

[6] R. Verhulst, "Op weg naar inzicht", *Wiskunde en Onderwijs* **20**:227-40, 1994.

rang M = 2
$= \text{rang } [M \vdots B] \Longleftrightarrow$ L snijdt Q

rang M = 1 en
- rang $[M \vdots B] = 1 \Longleftrightarrow L = Q$
- rang $[M \vdots B] = 2 \Longleftrightarrow L \cap Q = \emptyset$

2) in dimensie n > 2

$$\begin{cases} \alpha \leftrightarrow a_1x_1 + a_2x_2 + \ldots + a_nx_n = a_{n+1} \\ \beta \leftrightarrow b_1x_1 + b_2x_2 + \ldots + b_nx_n = b_{n+1} \end{cases}$$

$$\to \left[\begin{array}{ccc|c} a_1 & \ldots & a_n & a_{n+1} \\ b_1 & \ldots & b_n & b_{n+1} \end{array}\right] = [M \vdots B]$$

rang M = 2
$= \text{rang } [M \vdots B] \Longleftrightarrow \alpha$ snijdt β

rang M = 1 en
- rang $[M \vdots B] = 1 \Longleftrightarrow \alpha = \beta$
- rang $[M \vdots B] = 2 \Longleftrightarrow \alpha \cap \beta = \emptyset$

dan hangt $L \cap Q$ af van het aantal oplossingen voor de parameters r en s in het stelsel :

$$\begin{cases} x_1 + ra = x_2 + sc \\ y_1 + rb = y_2 + sd \end{cases}$$

of nog :

$$\begin{cases} ar - cs = x_2 - x_1 \\ br - ds = y_2 - y_1 \end{cases}$$

$$\to \left[\begin{array}{cc|c} a & -c & x_2 - x_1 \\ b & -d & y_2 - y_1 \end{array}\right] = [M \vdots B]$$

rang M = 2
$= \text{rang } [M \vdots B] \Longleftrightarrow$ L snijdt Q

rang M = 1 en
- rang $[M \vdots B] = 1 \Longleftrightarrow L = Q$
- rang $[M \vdots B] = 2 \Longleftrightarrow L \cap Q = \emptyset$

2) in dimensie n > 2

$L \leftrightarrow (x_1, x_2, \ldots, x_n) + R(a_1, a_2, \ldots, a_n)$
$Q \leftrightarrow (y_1, y_2, \ldots, y_n) + R(b_1, b_2, \ldots, b_n)$

$$\begin{cases} a_1r - b_1s = y_1 - x_1 \\ a_2r - b_2s = y_2 - x_2 \\ \ldots\ldots\ldots\ldots \\ a_nr - b_ns = y_n - x_n \end{cases}$$

$$\to \left[\begin{array}{cc|c} a_1 & -b_1 & y_1 - x_1 \\ a_2 & -b_2 & y_2 - x_2 \\ \ldots & \ldots & \ldots \\ \ldots & \ldots & \ldots \\ a_n & -b_n & y_n - x_n \end{array}\right] = [M \vdots B]$$

rang M = 2 en
- rang $[M \vdots B] = 2 \Longleftrightarrow$ L snijdt Q
- rang $[M \vdots B] = 3 \Longleftrightarrow$ L kruist Q

rang M = 1 en
- rang $[M \vdots B] = 1 \Longleftrightarrow L = Q$
- rang $[M \vdots B] = 2 \Longleftrightarrow L \,//\, Q$ en $L \cap Q = \emptyset$

Figure 4

In this example the possibilities of obtaining the intersection between the lines are compared.

And a final subcategory is the one in which historical and philosophical topics are transformed in such a way that they can be used in the classroom. In this kind of article a panoply of pedagogical techniques is used, ranging from anecdote to a survey of a particular topic. Since the history of mathematics holds virtually an infinity of anecdotes and/or techniques which can be adapted. The following example[7] (figure 5) shows how notation is important in mathematics. It uses the example of Diophantus to illustrate that the notation of a polynomial can be confusing without using + or – signs.

It also shows the Ionic numeral system as a transition between the additive and the decimal positional system. (The Ionic system is positional but it uses different symbols for 1 when the 1 indicates 1, 10 or 100, thus 111 would be written as υκα).

Hij gebruikte de volgende regels voor het schrijven van een veelterm :

1. De coëfficienten worden geschreven in Griekse cijfers na de onbekende.
2. Alle termen die afgetrokken dienen te worden worden geschreven na Λ.
3. De termen die men moet optellen, worden zonder somteken aan de ene kant geschreven, de termen die afgetrokken dienen te worden, aan de andere kant.

Bijvoorbeeld :

$\Delta^{Y}\bar{\delta}M^{0}\overline{\kappa\varepsilon} = 4x^2 + 25$;

$\Delta^{Y}\Delta\bar{\alpha}M^{0}\bar{\omega}\Lambda\Delta^{Y}\phi = x^4 - 50x^2 + 800$;

$K^{Y}\bar{\beta}\overline{\varsigma\eta}\Lambda\Delta^{Y}\bar{\varepsilon}M^{0}\bar{\alpha} = 2x^3 - 5x^2 + 8x - 1$.

4. Oefeningen

Schrijf in Ionische cijfers :

45 ;	109 ;
336 ;	8957 ;
$\frac{5}{3}$;	$\frac{921}{508}$.

Figure 5

Actually part of this lesson was not given in a mathematics class, but in a Greek class.

Historical examples are one way of tearing down barriers between subjects taught in a secondary school. These examples tend not only to have a relation with other scientific subjects, but also with language courses.

[7] N. Van der Auwera, "Diophantus in de klas", *Wiskunde en Onderwijs* **21**:207-12,1995.

4. ARTICLES RELATING TO NEW RESEARCH

We can be relatively brief about this category, for the title says it all. In the mid-nineties these would relate, e.g., to Wiles (Fermat) and number theory, while in the eighties they related to fractals. In this category, whether an article is received well depends on the ability of the author to translate his subject matter from a tale for specialists to one for a lay public (at least in his field of research). From time to time, scholars also submit articles actually containing new research. In general, these are refused, and in a letter to the author (s)he is referred to a more appropriate journal. Things are somewhat different however if the article comes not from a research scientist, but from a secondary school teacher, say an "amateur". In other sciences, the work of amateurs is readily acknowledged, but in mathematics it tends to be perceived with suspicion. Probably the reason for this is the many crackpots trying to prove the quadrature of the circle and the like.

At one point, *Wiskunde en Onderwijs* received short articles by Guido Lasters, concerning his "algebra with points". They faced the editors with a difficult task. The articles did not contain any mistakes, but they went off the beaten track, for reasons unapparent. The board finally decided to publish[8]. And something wonderful happened: the content of the articles was picked up by professor Fonny Ooms of the Limburg Universitair Centrum (LUC), who, together with Guido Lasters, forced the ideas into a strict theoretical framework[9]. Their joint work was called "Thales Spaces", and thus were born a new kind of space and a new kind of theory. The latter was built on three axioms for which the Lasters' articles had given special cases, or examples of existence. Figures 6 and 7 show extracts of Lasters' original manuscripts and the resulting paper with Ooms respectively.

[8] G. Lasters, "De stelling van Thales en het begrip distributiviteit", *Wiskunde en Onderwijs* **8**: 275-6, 1982. Also published as G. Lasters, " The distributive law in geometry", *Mathematical Spectrum* **18/2**: 41, 1985-86.

[9] F. Ooms, G. Lasters en L. Froidcoeur, "Thalesruimten", *Wiskunde en Onderwijs* **13**: 402-21, 1987.

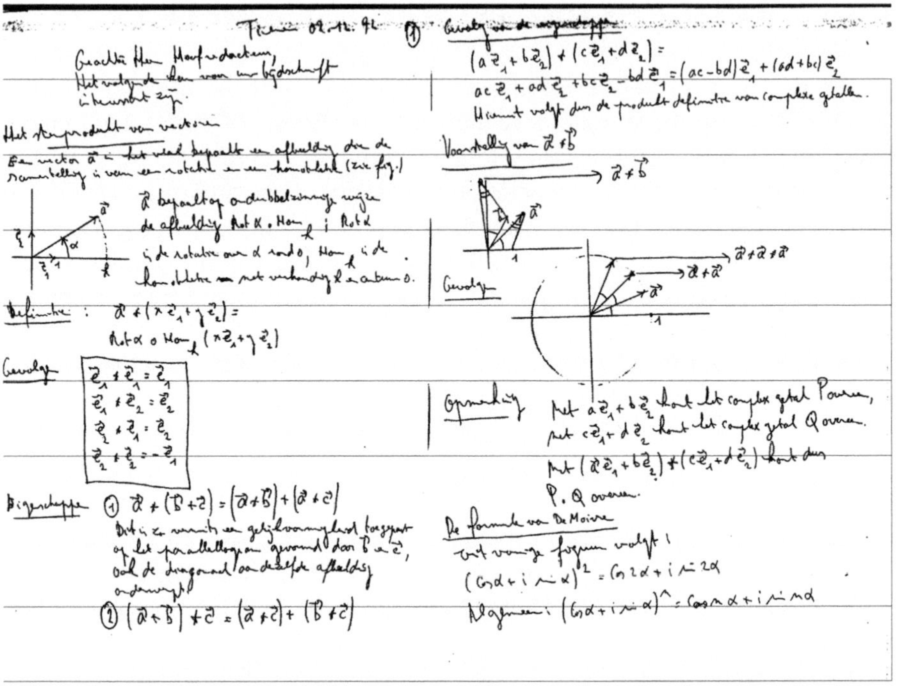

Figure 6

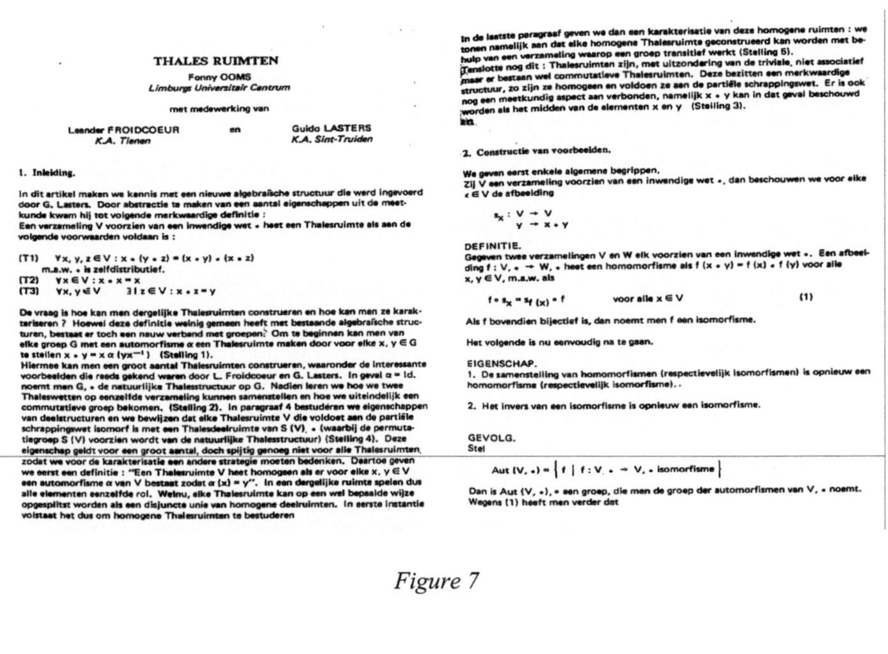

THALES RUIMTEN

Fonny OOMS
Limburgs Universitair Centrum

met medewerking van

Leander FROIDCOEUR *K.A. Tienen* en Guido LASTERS *K.A. Sint-Truiden*

1. Inleiding.

In dit artikel maken we kennis met een nieuwe algebraïsche structuur die werd ingevoerd door G. Lasters. Door abstractie te maken van een aantal eigenschappen uit de meetkunde kwam hij tot volgende merkwaardige definitie :
Een verzameling V voorzien van een inwendige wet $*$ heet een Thalesruimte als aan de volgende voorwaarden voldaan is :

(T1) $\forall x, y, z \in V : x * (y * z) = (x * y) * (x * z)$
m.a.w. $*$ is zelfdistributief.
(T2) $\forall x \in V : x * x = x$
(T3) $\forall x, y \in V \quad \exists! z \in V : x * z = y$

De vraag is hoe kan men dergelijke Thalesruimten construeren en hoe kan men ze karakteriseren ? Hoewel deze definitie weinig gemeen heeft met bestaande algebraïsche structuren, bestaat er toch een nauw verband met groepen. Om te beginnen kan men van elke groep G met een automorfisme α een Thalesruimte maken door voor elke $x, y \in G$ te stellen $x * y = x\,\alpha\,(yx^{-1})$ (Stelling 1).
Hiermee kan men een groot aantal Thalesruimten construeren, waaronder de interessante voorbeelden die reeds gekend waren door L. Froidcoeur en G. Lasters. In geval α = Id. noemt men G, $*$ de natuurlijke Thalesstructuur op G. Nadien leren we hoe we twee Thaleswetten op eenzelfde verzameling kunnen samenstellen en hoe we uiteindelijk een commutatieve groep bekomen. (Stelling 2). In paragraaf 4 bestudéren we eigenschappen van deelstructuren en we bewijzen dat elke Thalesruimte V die voldoet aan de partiële schrappingswet isomorf is met een Thalesdeelruimte van S (V), $*$ (waarbij de permutatiegroep S (V) voorzien wordt van de natuurlijke Thalesstructuur) (Stelling 4). Deze eigenschap geldt voor een groot aantal, doch spijtig genoeg niet voor alle Thalesruimten, zodat we voor de karakterisatie een andere strategie moeten bedenken. Daartoe geven we eerst een definitie : "Een Thalesruimte V heet homogeen als er voor elke $x, y \in V$ een automorfisme α van V bestaat zodat $\alpha(x) = y$". In een dergelijke ruimte spelen dus alle elementen eenzelfde rol. Welnu, elke Thalesruimte kan op een wel bepaalde wijze opgesplitst worden als een disjuncte unie van homogene deelruimten. In eerste instantie volstaat het dus om homogene Thalesruimten te bestuderen

In de laatste paragraaf geven we dan een karakterisatie van deze homogene ruimten : we tonen namelijk aan dat elke homogene Thalesruimte geconstrueerd kan worden met behulp van een verzameling waarop een groep transitief werkt (Stelling 5).
Tenslotte nog dit : Thalesruimten zijn, met uitzondering van de triviale, niet associatief maar er bestaan wel commutatieve Thalesruimten. Deze bezitten een merkwaardige structuur, zo zijn ze homogeen en voldoen ze aan de partiële schrappingswet. Er is ook nog een meetkundig aspect aan verbonden, namelijk $x * y$ kan in dat geval beschouwd worden als het midden van de elementen x en y (Stelling 3). [illegible].

2. Constructie van voorbeelden.

We geven eerst enkele algemene begrippen.
Zij V een verzameling voorzien van een inwendige wet $*$, dan beschouwen we voor elke $x \in V$ de afbeelding

$$s_x : V \to V, \quad y \to x * y$$

DEFINITIE.
Gegeven twee verzamelingen V en W elk voorzien van een inwendige wet $*$. Een afbeelding $f : V, * \to W, *$ heet een homomorfisme als $f(x * y) = f(x) * f(y)$ voor alle $x, y \in V$, m.a.w. als

$$f \circ s_x = s_{f(x)} \circ f \qquad \text{voor alle } x \in V \qquad (1)$$

Als f bovendien bijectief is, dan noemt men f een isomorfisme.

Het volgende is nu eenvoudig na te gaan.

EIGENSCHAP.
1. De samenstelling van homomorfismen (respectievelijk isomorfismen) is opnieuw een homomorfisme (respectievelijk isomorfisme).
2. Het invers van een isomorfisme is opnieuw een isomorfisme.

GEVOLG.
Stel

$$\text{Aut}(V, *) = \{ f \mid f : V, * \to V, * \text{ isomorfisme} \}$$

Dan is Aut (V, $*$), $\circ$ een groep, die men de groep der automorfismen van V, $*$ noemt.
Wegens (1) heeft men verder dat

Figure 7

5. PROBLEM CORNER AND RELATED ARTICLES

In our view, the most fundamental part of a journal for mathematics teachers is the problem corner. This part of the journal engages its readers in doing what they have studied for: solving specific and concrete mathematical problems. Although in the majority of cases the problems posed have already been solved, the importance lies in the challenge the readers face. This points to one of the essential dimensions of doing mathematics: the satisfaction one has when having solved a problem by him- or herself. It does not matter whether it has been solved before, *you* did it. The structure of DNA cannot be discovered over and over again, but Pythagoras' theorem *can* be proved over and over again. Indeed, *new* proofs for the theorem can be found. E.g., Verhulst's logistical equation (1849) was once 'solved' by a contributor to the problem corner who was completely unaware of the existing solutions. Unlike Verhulst, he approached the problem from the viewpoint of a computer scientist engaged with fractals[10].

In Figure 8 we see the general concept he used, a systematic search for properties of the sequence, complete with proofs and corollaries.

Although readers are invited to post their solutions to the problem corner, editor experience shows that only a minority do this. In contrast, the answers, published in a subsequent issue, are probably the most widely read part of the journal. There are two big subcategories in the problem corner. The first contains articles in which the author poses a question and then solves it. For instance, in *The Mathematical Gazette* Tom Roper once asked himself what is the path a bowl follows in the game of bowls (Figure 9)[11]. The other contains articles generated by taking interest in a problem posed in the corner.

The problem corner becomes even more interesting when readers themselves start to pose questions. One of those came from L. Huybrechts and seemed really very simple. But then, are the most simple of mathematical questions not the most interesting ones? Here it is: *Suppose you build piramids with oranges. You can build one with a triangular base and one with a square base. Can a triangular and a square piramid be built with the same number of oranges?*[12] The answer to the problem, hitherto unsolved (to the best of our knowledge), is to be found in the theory of Diophantine equations.

[10] In: R. Laumen, "Zoekersrubriek", *Wiskunde en Onderwijs* **14**: 89-107, 1988.

[11] Tom Roper, "The mathematics of bowls", *The Mathematical gazette* **80**: 298-307, 1996.

[12] Question in : R. Laumen, "Zoekersrubriek", *Wiskunde en Onderwijs* **12:** 175-87, 1986. 'Solutions' in R. Laumen, "Zoekersrubriek", *Wiskunde en Onderwijs* **12**, 424—35, 1986.

Zoekersbijdrage.

Dit is een nieuw onderdeel in onze zoekersrubriek en zal telkens aan bod komen wanneer een zoeker een vrij uitgebreide oplossing meebrengt maar wiskundig en didactisch interessant genoeg is voor publicatie.
Volgende zoeker - afkomstig van zoekersvriend CLAESEN - voldoet zonder twijfel aan vermelde criteria.
Opgave van de bewuste zoeker.

"Voor welke waarde van a $\in \mathbb{R}$ *convergeert de rij*
$x_0, x_1, \ldots, x_{n+1} = a \,.\, x_n\,(1 - x_n), \ldots$ "

Zoekerscollega en hoofdredacteur MESKENS zond hierover een merkwaardige bijdrage in. Ziehier zijn pennevrucht, door de redacteur van deze rubriek lichtelijk didactisch bewerkt.

I. ALGEMENE EIGENSCHAPPEN VAN DE OPGEGEVEN RIJ.

Eigenschap 1 : De rij verandert niet $(k > n)$ als men x_k door $1 - x_k$ vervangt.

Bewijs : Stel $\bar{x}_n = 1 - x_n$

$$\begin{aligned} \text{dan is } \bar{x}_{n+1} &= a\,\bar{x}_n\,(1 - \bar{x}_n) \\ &= a\,(1 - x_n)\,x_n \\ &= x_{n+1} \end{aligned}$$

Gevolg 1 : De rijen met beginterm x_0 en $1 - x_0$ hebben dezelfde convergentie-eigenschappen.

Figure 8

If the square piramid has m layers, and the triangular one $n = m + x$ the resulting equation, Jacques Rasking found, which has to be solved in the positive integers is: $(m + 1)(m + 1 + x)(m + 2 + x) = m(m + 1)(2m + 1)$. As an exercise for the readers of this article who have an interest in Diophantine equations, we can generalize the problem thus: *Suppose you build piramids with oranges. An n-piramid is a piramid with a regular n-gon as base. Do numbers m and n exist such that the number of oranges in the m-piramid is the same as the number in the n-piramid?*

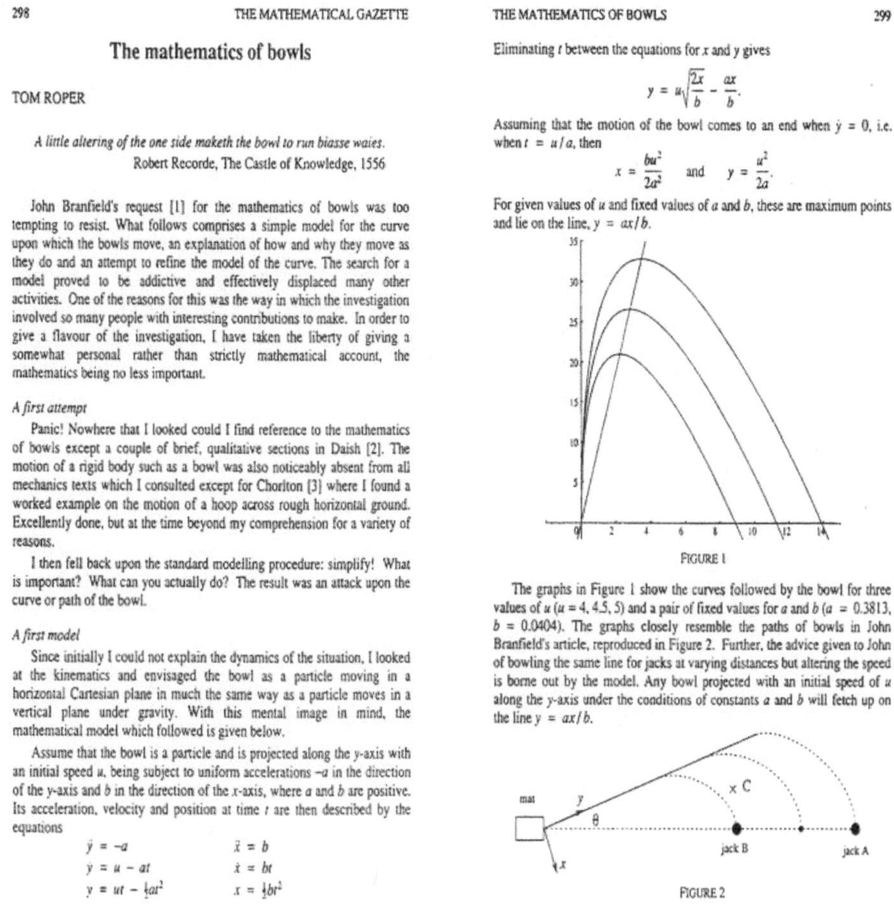
298 THE MATHEMATICAL GAZETTE

The mathematics of bowls

TOM ROPER

A little altering of the one side maketh the bowl to run biasse waies.
Robert Recorde, The Castle of Knowledge, 1556

John Branfield's request [1] for the mathematics of bowls was too tempting to resist. What follows comprises a simple model for the curve upon which the bowls move, an explanation of how and why they move as they do and an attempt to refine the model of the curve. The search for a model proved to be addictive and effectively displaced many other activities. One of the reasons for this was the way in which the investigation involved so many people with interesting contributions to make. In order to give a flavour of the investigation, I have taken the liberty of giving a somewhat personal rather than strictly mathematical account, the mathematics being no less important.

A first attempt

Panic! Nowhere that I looked could I find reference to the mathematics of bowls except a couple of brief, qualitative sections in Daish [2]. The motion of a rigid body such as a bowl was also noticeably absent from all mechanics texts which I consulted except for Chorlton [3] where I found a worked example on the motion of a hoop across rough horizontal ground. Excellently done, but at the time beyond my comprehension for a variety of reasons.

I then fell back upon the standard modelling procedure: simplify! What is important? What can you actually do? The result was an attack upon the curve or path of the bowl.

A first model

Since initially I could not explain the dynamics of the situation, I looked at the kinematics and envisaged the bowl as a particle moving in a horizontal Cartesian plane in much the same way as a particle moves in a vertical plane under gravity. With this mental image in mind, the mathematical model which followed is given below.

Assume that the bowl is a particle and is projected along the y-axis with an initial speed u, being subject to uniform accelerations $-a$ in the direction of the y-axis and b in the direction of the x-axis, where a and b are positive. Its acceleration, velocity and position at time t are then described by the equations

$$\ddot{y} = -a \qquad \ddot{x} = b$$
$$\dot{y} = u - at \qquad \dot{x} = bt$$
$$y = ut - \tfrac{1}{2}at^2 \qquad x = \tfrac{1}{2}bt^2$$

THE MATHEMATICS OF BOWLS 299

Eliminating t between the equations for x and y gives

$$y = u\sqrt{\frac{2x}{b}} - \frac{ax}{b}.$$

Assuming that the motion of the bowl comes to an end when $\dot{y} = 0$, i.e. when $t = u/a$, then

$$x = \frac{bu^2}{2a^2} \quad \text{and} \quad y = \frac{u^2}{2a}.$$

For given values of u and fixed values of a and b, these are maximum points and lie on the line, $y = ax/b$.

FIGURE 1

The graphs in Figure 1 show the curves followed by the bowl for three values of u (u = 4, 4.5, 5) and a pair of fixed values for a and b (a = 0.3813, b = 0.0404). The graphs closely resemble the paths of bowls in John Branfield's article, reproduced in Figure 2. Further, the advice given to John of bowling the same line for jacks at varying distances but altering the speed is borne out by the model. Any bowl projected with an initial speed of u along the y-axis under the conditions of constants a and b will fetch up on the line $y = ax/b$.

FIGURE 2

Figure 9

The point of the question is clear enough: will it generate difficult Diophantine equations or on the contrary easy ones? We do not know. The only point we are trying to make is that many problems in the history of mathematics have indeed arisen from these 'simple' questions. Fermat's Last Theorem, Kepler's Packing Problem, the four colour problem – which incidently was posed by a student to Augustus de Morgan, who then forwarded it. Problem corners in mathematics teachers journals will always generate this kind of questions. Some may be picked up by research scientists, some may be forgotten for decades and then spring to life again. Some may even be posed, simultaneously or not, in many journals all over the world. Problem corners are the go-between of research mathematicians and mathematics teachers.

6. CONCLUSION

We hope the above paragraphs have shown that a journal for mathematics teachers is very necessary, not only for the teachers, but for the mathematical community at large. It is not only important for disseminating new ways of teaching, but also for keeping the mathematical mind occupied with that at which it is best: problem solving. Only in this way a vast, nearly untapped, reservoir of mathematical minds can continue to generate new problems and, with the aid of the research community, new answers. It is one of the distinguishing elements which separates mathematics from the other sciences: crackpots apart, there are no real amateurs in mathematics.

Chapter 13

ON THE INTERDISCIPLINARY STUDY OF MATHEMATICAL PRACTICE, WITH A REAL LIVE CASE STUDY

Reuben Hersh
University of New Mexico

Abstract: The study of mathematical practice is not the private property of philosophers. Many other disciplines, especially including mathematicians, are engaged in it. As a case study, an elementary derivation of Heon's area formula is presented, and then analyzed methodologically and ontologically.

Key words: Mathematical practice, Heron's formula, tetrahedron, mathematical existence

Of course, various special sorts of math studies have been going on for a long time. Philosophical, since Pythagoras/Plato. Historical, since Proclus. Anthropological, sociological and psychological, somewhat more recently. Now starting to see neurophysiological studies of math. Sometimes the different specialties have been fiercely autonomous. Gottlob Frege (to whom all contemporary philosophy of math is said to be but a footnote) banned psychologizing and historicizing from the philosophy of mathematics. On the other hand, André Weil, the great number theorist, entered the history of math with a dictum: no two disciplines have less in common than the history of mathematics and the philosophy of mathematics.

The conference from which this paper originates was called by philosophers. Two mathematicians were on the program, and there were mathematics educators, and a sociologist, and some historians. I hope this interdisciplinary approach will be continued and extended. Imre Lakatos (paraphrasing Immanuel Kant) said, as I remember, "Without history, the philosophy of mathematics is sterile; without philosophy, the history of mathematics is aimless." Yet I hear quiet intimations that the participation of mathematicians themselves is – shall I say – unexpected? This is a lamentable misapprehension! We mathematicians have a special, direct,

B. Van Kerkhove and J.P. van Bendegem (eds.), Perspectives on Mathematical Practices, 231–241.

intimate acquaintance with mathematical practice. Some of us have written about our practice.

Everyone knows George Polya's famous books on mathematical heuristics – a salient aspect of mathematical practice. The articles and books of Raymond Wilder are not obscure. He was a leading mathematician who applied the viewpoint of anthropology to his own field. There are the famous writings of Henri Poincaré on the psychology of mathematical discovery. And the famous book on that subject by his student Jacques Hadamard. The memoirs of Gian-Carlo Rota, Alfréd Rényi, John Littlewood, G. H. Hardy, Norbert Wiener, Paul Halmos, Stanislaus Ulam, Mark Kac, Paul Lévy, Laurent Schwartz, are rich with comments and revelations on mathematical practice, as are the two volumes of interviews, "Mathematical People" and "More Mathematical People." It is true that mathematicians' talk about mathematical practice is not in the logicist style inherited from Russell, Carnap and Quine, nor is it ornamented with statistically massaged "quantitative data" in the prevalent style of Anglophone psychology or sociology. But surely such shortcomings can be tolerated.

Enough querulous comments. I now turn to a more constructive tone. What would be the goal of a study of mathematical practice? What questions would one hope to answer, or at least clarify?

Here are some obvious questions that are subject to standard research techniques: Who are these mathematicians, sociologically, psychologically, anthropologically? How are their activities organized—institutionally, politically, economically? What are their attitudes and emotions about mathematical activity? Etc.

These are questions one asks about any sect or clique. But the point of studying mathematical practice is to understand its uniqueness. Its special character that makes it what it is. The invisibility, intangibility, of its objects of study. The unique consensus or "certainty" of its "results." And its "unreasonable" effectiveness in understanding and mastering Nature. These are the big philosophical questions. It's not clear how studying mathematical practice can answer them.

When Carnap or Quine philosophized about mathematics, they philosophized about what analytic philosophy postulated mathematics to be. Not about what mathematicians actually do. Why then do some philosophers now think that mathematical practice is philosophically interesting? Lakatos exposed this scandal. Philip Kitcher made an impressive effort to give a philosophical description of mathematical practice. Perhaps the impulse for philosophers to study mathematical practice now is their response to the exposure of the irrelevance of their predecessors.

As a mathematician invited to this project, I can contribute a case study from my own mathematical practice, presented in as philosophical a style as

I can summon up. It requires only the concept of area of a triangle, and 11th-grade algebra. (referring to the U.S. high school.) "Heron's area formula" is usually omitted from standard high school math, even though it is very simple and easy to use. The classical proof is long and tricky. I found a very simple proof of this formula. I here offer it as a case study in mathematical practice.

Here is Heron's formula:

> If a, b, c are the lengths of the sides, and s, (the "semi-perimeter") = (a+b+c)/2, then the area is $\sqrt{s(s-a)(s-b)(s-c)}$.

> (For example, take the right triangle with sides 3,4,5. By the elementary formula "area = ½ base times altitude," it has area ½ .3.4 = 6. The semi-perimeter is 6. Heron gives $\sqrt{6.1.3.2}$ which of course checks.)

The excellent book "Journey Through Genius" by William Dunham takes seven pages for the proof. Dunham comments, "This is a very peculiar result, which, at first glance, looks like nothing if not a misprint. The presence of the square root and semiperimeter seems odd, and the formula has no intuitive appeal whatever. The proof that Heron furnished is at once extremely circuitous, extremely surprising, and extremely ingenious."

I will go through my thoughts and afterthoughts, in 40 very short steps. Then I will look back, and look for some general conclusions.

1. The first step was reading the classical proof in Dunham's fine book and being bothered that while Heron's formula is so simple, its classical proof is so complicated. Such a simple formula "must" have a simple, natural proof.

2. The next step is to formulate this "feeling" as a problem: Find a simple, natural proof of Heron's area formula. This is a short step, but it is essential. Only this second step leads to action.

3. Stare at the formula. Look for something to do with it or to it.

4. Notice that the letter "s", the semi-perimeter, can be eliminated. (In Heron's proof, the semi-perimeter plays a starring role.)

5. A purely "mechanical" step: substitute the definition of "s" into Heron's formula, and combine terms. You get a "new" formula:

$$\text{Area} = \sqrt{\{(a+b+c)/2\}\{(-a+b+c)/2\}\{(a-b+c)/2\}\{(a+b-c)/2\}}$$

6. This formula makes even clearer the symmetry between the three letters a,b,c.

7. Under the square root sign we have a *polynomial.* In fact, a symmetric polynomial in three variables a,b,c. This simple remark is the key insight! It makes the crucial connection with Algebra!

8. As soon as a polynomial is noticed, I think of finding its roots-- finding the values of a,b and c that make it equal zero. Why? Because solving polynomials-finding their roots-is what one usually has to do to a polynomial. This polynomial is already factored! I see immediately that it equals zero if and only if either (a+b+c) or (-a+b+c) or (a-b+c) or (a+b-c) equals zero.

9. But what does that mean geometrically? What would that say about the triangle?

10. Since the side lengths a,b,c all are positive, a+b+c cannot equal zero.

11. If -a+b+c =0, then a = b+c.

12. That says that two side lengths are equal to the third. The triangle is "degenerate." Side "a" lies on top of sides "b" and "c". The area is zero.

13. Of course! If the triangle is degenerate, the area MUST equal zero! The formula for the area must vanish to zero when a = b + c.

14. But that means that -a+b+c must be a factor in the area formula. (This is the "Factor Theorem", from the 11th grade.)

15. There is nothing special about "a" among the three letters. Therefore, along with (-a+b+c), (a-b+c) and (a+b-c) must also be factors, This interchangeability of the letters is called "symmetry."

16. With at least three distinct factors, the area as a function of the side lengths would have to have degree at least three.

17. But area scales like a quadratic, a second power of lengths. I could get a quadratic function by taking the two-thirds power of a cubic. But that seems

ugly and unnatural. It is much more "natural" to look for one more first-degree ("linear") factor, and then take the square root of the resulting fourth-degree (quartic) expression.

18. What could the fourth linear factor be? It would have to preserve symmetry in the three letters a,b,c. So it could only be a linear combination of them with equal coefficients. I may as well take those coefficients to be 1.

19. So the vanishing of the degenerate cases and symmetry give me all the factors of the area function, assuming the area is the square root of a quartic polynomial. I still have to determine an arbitrary constant factor of multiplication.

20. To find that constant, I can use any triangle of known area. If a=3, b=4, c=5, area = 6, I find that $\sqrt{(a+b+c)(-a+b+c)(a-b+c)(a+b-c)}$ is $\sqrt{12.6.4.2}$ which is 24. So I must multiply by ¼ to get the area, 6.

21. I have derived an area formula:

$$\text{Area} = \tfrac{1}{4}\sqrt{(a+b+c)(-a+b+c)(a-b+c)(a+b-c)}$$

which is the same as Heron. (In Heron, each factor under the square root sign is divided by 2, which gives a denominator of 16 under the square root sign. If you take the denominator outside the square root sign, you must take the square root of 1/16, and get ¼.)

22. Am I done? I have obtained a quick derivation of Heron's formula, using: the symmetry of the three letters a,b,c, the requirement that that the area equal zero when the triangle is degenerate, the quadratic dependence of area on length, the area for the special triangle with sides 3,4,5, AND the assumption that the area is the square root of a quartic. But what right do I have to assume THAT?

23. The area formula must contain at least three independent linear factors, so it can't just be a quadratic polynomial. Could it be the cube root of a sextic, or the fourth root of an octic?

24. To eliminate these possibilities, I need to verify this formula-prove it.

My derivation, which leads me to find the formula, is not a proof that the formula works correctly in all cases.

25. In preparation for a proof, I multiply out the four linear factors. The product will be of fourth degree, symmetric in the three variables a,b,c. Furthermore, replacing a,b,or c by its negative leaves the product unchanged, so the product is an "even" function. It can contain only even powers, no first or third powers.

26. Consequently, the resulting multiplied-out quartic can only contain 4th powers, products of two squares, and possibly an additive constant.

27. Because the area is zero if any of the sides is zero, the additive constant must equal zero. So the multiplied-out form, by symmetry, must be a linear combination of two expressions: the sum of three fourth powers, and the sum of three products of two squares.

28. To find the two unknown coefficients, I check with two specific triangles, and obtain

$$\text{area} = \tfrac{1}{4}\sqrt{2(a^2b^2 + b^2c^2 + a^2c^2) - (a^4 + b^4 + c^4)}$$

Of course this is the same thing I would have obtained by multiplying out from the factored form.

29. How can I verify this formula – prove it is valid for *all* triangles?

30. I know a standard trick – introduce x-y coordinates.

31. For maximum simplicity of algebra, I put the origin at a vertex of the triangle, with the x-axis on one side of the triangle, and the point $x = 1$, $y=0$ at a second vertex. Then I choose the unit of length on the y-axis equal to the corresponding altitude. The third vertex is then at $(x,1)$ for some value of "x".

32. The base of the triangle has length 1, and the altitude has length 1, so the area in these coordinates is 1/2, independent of "x". The side lengths squared are now $a^2 = 1$, $b^2 = x^2 + 1$, and $c^2 = (x-1)^2 + 1$.

33. Plugging these expressions into my multiplied-out area formula, the

variable "x" cancels out, the algebra simplifies to the correct answer, ½, and Heron's formula is verified.

34. FINISHED! I have derived and then proved Heron's formula, without any ingenious geometrical constructions, and without trigonometry!! (In fact, from this independent proof of Heron, I can get an independent derivation of the law of cosines. This is a generalization of the Pythagorean theorem, which Dunham laboriously derives from Heron.)

35. What next? I won't stop now, after only one victory. What new problem is now inviting me?

36. Is there a Heron's formula for the non-Euclidean triangle? Is there a Heron's formula for the Euclidean quadrilateral? Is there a Heron's formula in three dimensions, for the volume of a tetrahedron?

37. I checked with three friends (an editor of the "College Mathematics Journal", a colleague who is a well-known expert on problem-solving, and my respected mentor Peter Lax.) None of them had seen my proof before. All encouraged publication. Professor Lax sent me a strangely complicated yet somehow elegant formula for the volume of a tetrahedron.

38. "Focus", a publication of the *Mathematical Association of America*, accepted my article [November, 2002, Volume 22, Number 8, page 22].

39. The editor of "Focus" [January, 2003, Volume 23, Number 1, page 15] received letters from readers pointing out that my derivation of Heron's formula had already appeared in 1987, in a note in the "College Mathematics Journal". Moreover, my formula for the volume of a certain tetrahedron was wrong, the denominator should be 12, not 6. Moreover, a formula for the volume of a tetrahedron as the square root of a sixth degree polynomial in the lengths of the sides is already in the literature, in George Polya's "Mathematics and Plausible Reasoning." (While writing this article, I found a reference [Dorrie, p. 285] stating that Euler had published this tetrahedron formula in 1753.)

40. I presented this story in Brussels at the meeting, "Perspectives on Mathematical Practices." The night before my lecture, I checked Lax's formula for the volume of a tetrahedron. It is wrong! (It turned out that he had forgotten a factor of 1/6.) Polya gives Euler's tetrahedron formula without a derivation or proof. Lax's derivation of it generalizes to higher dimensions. But it is unknown whether there is a factorization, even in the

three-dimensional case given by Polya. There is a Heron-type formula for the area of a cyclic quadrilateral (one inscribed in a circle.) This is "Brahmagupta's formula." Maybe it wouldn't be too difficult to prove Brahmagupta by some adaptation of our proof of Heron.

This detailed account of a small piece of research in elementary mathematics is offered as a case study, accessible with minimal mathematical preparation. Of course it does not claim to be typical or representative. Other stories of mathematical investigation could be told. This one is not out of the ordinary. It may suggest some hypotheses of a methodological or even ontological kind.

So, "What is going on here?"

First of all, the original motivation is "esthetic." The classical proof seemed "too complicated" for such a simple result. There "must be" a simple proof.

Secondly, the known result is the starting point. I already know Heron's formula is true. But I want a simpler explanation, I want to understand it more directly.

Thirdly, the work is done against my whole background of mathematical know-how. Basic facts about triangles, areas, and polynomials are assumed and used. Standard arguments and manipulations are stored in my head, ready to be called on.

Fourthly, the key observation is this: three of the factors under the square root in Heron's formula are seen to be *necessary*. This insight makes the formula natural and understandable. The mystery evaporates.

Fifthly, it is a combination of two qualitative properties of area (symmetry and quadratic scaling) and a single quantitative one (vanishing of area in the degenerate cases) that *suggests* the form of the area function.

Sixthly, to find the final form of the formula, it is necessary to work out one special case numerically.

Seventhly, although the completed derivation is still not a proof, once the formula has been derived, the "rigorous" or complete proof is almost routine. Some insight was needed to *discover* the formula. Once it is known, the proof is not much more than an afterthought. Once the relationship between side lengths and area is understood, the details are easy.

These remarks suggest some properties of the part of mathematical practice that is called "research": the discovery of new results. An interesting new proof of an old result counts as a kind of new result. Research can start by noticing something "funny" – an unexpected or unexplained analogy. A peculiar complication that intuitively seems unnecessary. One's sense of fitness, of what feels right, can tell one where to look for something interesting.

One may not know in advance which tools, what background knowledge or know-how may come in handy as the investigation proceeds. The researcher expects to use his whole background of mathematical knowledge. Indeed, he may sometimes use analogies from outside of mathematics (perhaps from mechanics) and then his knowledge of mechanics may advance his investigation. He may need to go beyond his available skills, by going to the library or onto the Web or by calling up a friend.

The key event in the investigation may be making a fruitful connection between two different theorems or theories, two different ideas which click together. In this case, the Factor Theorem from elementary algebra, and the vanishing of the area of a degenerate triangle.

What is interesting to me may have been interesting to someone else. Once the question is asked, the answer is "there," waiting to be found by anyone who looks hard enough. In other words, the concept we are investigating already contains the answer hidden away. It is "pregnant" with the solution. So it is normal for the same discovery to be made several times.

The investigation may lead to a formula or "theorem" (precise statement) which is convincing on the basis of the derivation. Yet there may remain some question of the precise conditions under which it is true. There may be unnecessary hypotheses or assumptions used in the derivation. (In our case, that the area is the square root of a quartic; in the final proof, this is part of the conclusion.) Consequently, there may be a need for an a posteriori proof of the result. Without such a proof, it may be called a mere conjecture. Yet the derivation which led to the conjecture may leave no doubt that it is true, at least in certain important cases. The proof sometimes may be routine, compared to the derivation.

This last fact has important consequences for exposition and teaching. To present the proof without the derivation would then be a piece of mystification. The derivation would really be the heart of the matter, even though, from a pedantically "logical" point of view, the "rigorous" proof makes the original derivation superfluous. The slogan, "Mathematics is nothing without proof" becomes false when it degenerates to, "Mathematics is nothing but proof." Our example shows that sometimes proof is a less interesting and important part of the work. Finally, can we ascend from methodology to ontology? In other words, now that we have seen what is actually being done, does that tell us anything about the significance, the meaning, the import of all this conversation?

The most important thing that is forced on us is that there are facts of the matter. The area of a triangle may be defined in one way or another, but all the definitions have to match. Whatever definition you use, the area is one half the base times the altitude. (Any of the three bases, times the corresponding altitude.) And of course, Heron.

What then are we to make of the contemporary argument, whether mathematical objects are "real" or "fictional"? This is a perfectly useless question, unless some clarity is achieved on the meanings of "real" and "fictional". One could say, "Something is real if there are verifiably true statements about it. If we can have true knowledge of it, we can say it is real." If that is what we mean by "real", the area of a Euclidean triangle is some sort of real entity. If by "fictional" we mean arbitrary, something which is invented at the pleasure of the inventor (Mark Twain could have called Huck Finn by any other name, and changed his story any way he liked, subject only to his own fictional goals and tastes) then the area of a triangle is not a fiction. On the other hand, if by "real" you mean only an object that is made of atoms and molecules, then area is not real, and neither is "triangle" or "polynomial" or anything else we talked about in this whole tragedy. So what's needed, really, is a critical consideration of what we want to mean by "real", or, equivalently, what we want to mean by "exist".

This much can be said. Mathematics really exists. It is going on, it is taking place, it has been around a long time and is here to stay. If your vocabulary insists that it is not real, and since in any ordinary meaning of the word it is not "fictional", then you must find some other kind of ontology, neither "real" in your sense nor fictional in any sense, to place it in. My own answer to this conundrum is presented in my book.

REFERENCES

Albers, Donald J., and Gerald L. Alexanderson. "Mathematical People". Birkhauser, Boston, 1985.

Albers, Donald J. , Gerald L. Alexanderson and Constance Reid. "More Mathematical People." Birkhauser, Boston, 1990.

Alperin, C. "Heron's Area Formula," The College Mathematics Journal, March 1987, v. 18, no.2, pp. 137-8.

Arnold.V. I. "Huygens and Barrow, Newton and Hooke." Birkhauser, Boston, 1990.

Arnold, V. et al., editors. "Mathematics: Frontiers and Perspectives." International Mathematical Union and American Mathematicsl Society, 2000.

Crowe, M. "Ten "laws" Concerning Patterns of Change in the History of Mathematics. Historia Mathematica, 2, 161-166 (1975)

Davis, P.J. "Fidelity in Mathematical Discourse: Is 1+1 Really 2?" American Mathematical Monthly 78, 252-263 (1972)

Dehaene, Stanislas. "The Number Sense". Oxford University Press, New York, 1997.

Dieudonné, J. "The Work of Nicholas Bourbaki." American Mathematical Monthly. 77, 134-145 (1970)

Dorrie, Heinrich. "100 Great Problems of Elementary Mathematics," Dover Publications, New York, 1965.

Dunham, William. "Journey through Genius," Penguin Books, New York, 1990.

Hadamard, J. "The Psychology of Invention in the Mathematical Field." Dover Publications, New York, 1945.
Halmos, P. "Mathematics as a Creative Art." American Scientist. 56, 375-389 (1968)
Hardy, G. H. "Mathematical Proof." Mind. 38, 1-25 (1929)
Hersh, R. "What is Mathematics, Really?" Oxford University Press, New York, 1997.
Hersh, R. "A Nifty Derivation of Heron's Area Formula by 11th Grade Algebra." Focus, November 2002, v. 22, no.8, p. 22
Kac, M. "Enigmas of Chance" Harper and Row, New York, 1985.
Kitcher, P. "The Nature of Mathematical Knowledge." Oxford University Press, 1983.
Lakatos, I. "Proofs and Refutations." Cambridge University Pres, 1976.
Lévy, P. "Quelques Aspects de la Pensée d'un Mathematicien." Albert Blanchard 1970.
Littlewood, J. E. "Littlewood's Miscellany." Cambridge University Press, 1986.
MacLane, S. "Mathematics: Form and Function": New York. Springer-Verlag, 1986.
Mozzochi, . J. "The Fermat Diary." American Mathematical Society, 2000.
Peirce, C. S. "The Essence of Mathematics," In "Essays in the Philosophy of Science", Indianapolis: Bobbs-Merrill, 1957.
Pier, Jean-Paul, editor. "Development of Mathematics, 1950-2000." Birkhauser Boston, 2000.
Poincaré, Henri. "Mathematical Creation" in "The Foundations of Science". The Science Press, New York, 1913.
Poincaré, Henri. "The Nature of Mathematical Reasoning." In "Science and Hypothesis." Dover Publications, New York, 1952.
Polya, G. "How to Solve It." Princeton University Pres, 1971.
Polya, G. "Mathematical Discovery." John Wiley & Sons, 1971.
Polya, G. "Mathematics and Plausible Reasoning." Princeton University Press, 1954.
Reid, C. "Hlbert." Springer, 1972.
Reid, C. "Courant in Gottingen and New York." Springer, 1976.
Rényi, A. "Dialogues on Mathematics." San Francisco: Holden-Day, 1967.
Rota, Gian-Carlo. "Indiscrete Thoughts." Birkhauser, Boston, 1997.
Schwartz, L. "A Mathematician Grappling With His Century." Birkhauser, 2001.
Ulam, S. "Adventures of a Mathematician." Charles Scribner's Sons, 1983.
Von Neumann, J. "The Mathematician" In: "Works of the Mind," Robert B. Heywood (ed.) Chicago, University of Chicago Press, 1947.
Wiener, Norbert. "Ex-Prodigy." M.I.T. Press. 1964.
Wiener, Norbert. "I am a Mathematician." M.I.T. Press, 1956.
Wilder, Raymond L. "The Evolution of Mathematical Concepts." New York, John Wiley & Sons. 1968.
Wilder, Raymond L. "Hereditary Stress as a Cultural Force in Mathematics." Historia Mathematica, 1, 29-46 (1974)
Wilder, Raymond L. "Mathematics as a Cultural System," New York. Pergamon, 1981.
Yandell, Benjamin H. "The Master Class." A.K. Peters, 2002.

MIX
Papier aus verantwortungsvollen Quellen
Paper from responsible sources
FSC® C105338

If you have any concerns about our products,
you can contact us on
ProductSafety@springernature.com

In case Publisher is established outside the EU,
the EU authorized representative is:
Springer Nature Customer Service Center GmbH
Europaplatz 3, 69115 Heidelberg, Germany

Printed by Libri Plureos GmbH
in Hamburg, Germany